The Templeton Science *and* Religion Reader

To Doug
Treu

from

Emmanuel

Templeton Science and Religion Series

In our fast-paced and high-tech era, when visual information seems so dominant, the need for short and compelling books has increased. This conciseness and convenience is the goal of the Templeton Science and Religion Series. We have commissioned scientists in a range of fields to distill their experience and knowledge into a brief tour of their specialties. They are writing for a general audience, readers with interests in the sciences or the humanities, which includes religion and theology. The relationship between science and religion has been likened to four types of doorways. The first two enter a realm of "conflict" or "separation" between these two views of life and the world. The next two doorways, however, open to a world of "interaction" or "harmony" between science and religion. We have asked our authors to enter these latter doorways to judge the possibilities. They begin with their sciences and, in aiming to address religion, return with a wide variety of critical viewpoints. We hope these short books open intellectual doors of every kind to readers of all backgrounds.

Series Editors: J. Wentzel van Huyssteen & Khalil Chamcham
Project Editor: Larry Witham

The Templeton Science *and* Religion Reader

Edited by J. Wentzel van Huyssteen
and Khalil Chamcham

TEMPLETON PRESS

Templeton Press
300 Conshohocken State Road, Suite 500
West Conshohocken, PA 19428
www.templetonpress.org

Designed and typeset by Gopa and Ted2, Inc.

Library of Congress Cataloging-in-Publication Data

The Templeton science and religion reader /
edited by J. Wentzel van Huyssteen and Khalil Chamcham.
    p. cm.
Includes bibliographical references and index.
ISBN 978-1-59947-393-2 (pbk. : alk. paper) —
ISBN 978-1-59947-418-2 (e-book) 1. Religion and science.
I. Van Huyssteen, J. Wentzel (Jacobus Wentzel), 1942-
II. Chamcham, Khalil.
BL240.3.T44 2012
201'.65—dc23
                                        2012010209

Printed in the United States of America

12 13 14 15 16 17   10 9 8 7 6 5 4 3 2 1

Contents

The Templeton Science *and* Religion Reader

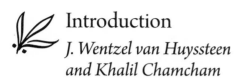 Introduction
J. Wentzel van Huyssteen
and Khalil Chamcham

THE INTEREST in the relationship between science and religion, both in academia and in the public, has grown in the past few decades. It has been driven by new findings in science and our perennial search for meaning through religion. This burst of interest is one of the most fascinating cultural phenomena of our time, but we would exaggerate to say that this never happened before. The volatile relationship between science and religion, and between specific sciences and distinct theologies, is centuries old. Our advantage today is to more fully understand how the boundaries between science and religion have shifted over the ages.

On that foundation, we have a fresh start. This collection of nine essays hopes to bring readers up to date on some of the great themes raised by the interdisciplinary encounter between different sciences and religions. Our contributors come from nine different fields of science. While they speak mostly about cutting-edge science, they also touch on the implications for religion. The essays are selected from the series of nine books that make up the Templeton Science and Religion Series, a project that unfolded between 2005 and 2011. Fortunately, the authors have given us the material necessary to take a panoramic view of the scientific findings that are relevant to religion, from the beginning of the universe to the everyday questions of the new computer technology and modern health care.

Science and religion are two of the most enduring, meaningful, and dominant cultural achievements of our species. They are both products of a remarkable historical development intimately inter-woven with the process of evolution, both biological and cultural. They are both part of the amazing cognitive story of our life on this Earth. The following selections illustrate how the interdisciplinary conversation between the different sciences and all aspects of the different religions is necessary for any comprehensive approach to human knowledge.

This is a very positive quest, but it also is challenging. Naturally, scientists are often reluctant to write on topics outside their own disciplines, and the topic of religion would seem to be very outside the box. Some scientists are highly literate in religion, theology, or the related history, but to master both fields is surely a rare feat even today. We were fortunate in assembling such a notable group in the Templeton Science and Religion Series. But the project also shows that there is plenty of room for more specialists in science and reli-gion, plenty of cultural barriers to cross, and a wide range of topics still to be researched with academic precision.

Why, after all, should we bring together two such different domains of knowledge? Some religious believers see science as a threat to their beliefs. Many scientists view religion as an insult to their rationality. As this collection of essays may suggest, the dynamic encounter of these two realms nevertheless leads us to more fundamental questions of concern to all sides—questions such as: What is knowledge? What is faith? What might they have in common? This current interest in science and religion has many names. Some call it a "dialogue," and others a "debate." We prefer to suggest it is a discourse, which has a much more open-ended feel-ing. It implies a dynamic and complex playing field with more than just two parties involved.

This volume is dedicated primarily to the facts and stories of sci-ence. The study of science, however, does not demand that it be taken as a "great alternative" to religious belief, as if a completely rational alternative is displacing an irrational one. Both of these kinds of knowledge are human. They are more complex and subtle than any dichotomy can contain. The dichotomy could be traced to the Renaissance and Enlightenment periods, a time when Western thought began to negate any previous modes of human logic and belief, whether the ancient religions or the disciplines such as astrology, alchemy, or mysticism. More often than not, science and religion today still are presented as two monolithic blocks of unchanging nature, different in their methodologies and separated by well-defined boundaries. True enough, religion is primarily about belief, prayer, and meditation, and science is about prediction and empirical proof.

On another view, however, the picture is more complex than that. On closer inspection, science is not monolithic at all. It is nearly impossible to generalize for all sciences. The most reliable way to approach science is through its specific fields, whether biology or quantum physics. Otherwise it is very easy to fall into vague generalization when trying to discuss "science." Each field of science has its own subject matter, history, methodology, theoretical construct, and experimental protocol. The attempt to define "science" continues in some quarters, but in the encounter between science and religion, it has proved important to take each field of science on its own terms.

The religion side of the coin is no less problematic. For one thing, in the distant past, the intellectual ferment within religion itself often gave rise to scientific approaches. Some aspects of religion have certainly contributed to the foundations of reason that now are expressed in science, law, politics, and ethics. At the same time, there are many kinds of religion, each with its unique by-products. The elements of science have been approached differently by the Jewish, Christian, Islamic, and Asian traditions. Religions also have engaged in their own internal debates, which has generated the field of theology. Different theologies have handled science in a variety of ways. Given this multiplicity of elements, it has become clear that the productive encounter between science and religion must be based on specific parts of each—a distinct scientific field and a precise theological tradition, for example—looking at the common issues they might share.

It is hard to deny that scientific progress usually takes priority even in an open discourse with religion. Nonetheless, science is far from having resolved the deep human anxieties about our origin and destiny. Science still needs help in this. In human society, science cannot hope to advance without paying attention to religious sensitivities. The debates within academic science and religion can be highly refined and extremely agitated. They should never lose sight, however, of the fact that the general public loses out if it cannot be presented with the new findings of science with an adequate explanation of the implication for society and religion. There is a time for heated debate, and a time for amicable analysis.

One way to avoid the simple clash over science and religion is to gain perspective. This is where historical scholarship has been helpful, revealing the "tidal motion" between science and religion as they developed into very specific forms. History is an important laboratory for seeing the many ways that science and religion have already had important encounters, and makes clear what remains to be done, or what seems to be a perennial concern. The historical context can reveal the important question of "epistemology"— the study of how we "know"—since this often is at the heart of the claims made by a particular science or a specific religion.

"Epistemology" is not everyone's favorite word, and it is one of the most complex aspects of philosophy, psychology, and even neuroscience. However, epistemology is no small matter. When science reaches a limit in its knowledge or applicability, the framework of epistemology must come to its rescue. The scientific field seriously grapples with how it knows things, either by theory, experiments, or mathematics. However, when science faces a new epistemological crisis—as it often does in history—it is not the role of religion to interfere. For example, when the now-conventional theory of the big bang origin of the universe was hotly debated in the mid-twentieth century, it took decades more of experiments to help settle the core issues. The attempt of some theologians to point out that the big bang had consonance with ideas of creation may have been meaningful, but it did not help at all in solving the scientific issues. Many other cases since then have proved that the best common ground between theology and new scientific discoveries has been a mutual reflection on epistemology, or the different ways of knowing, whether proposed by the priest or the scientist.

Nevertheless, the temptation to force one on the other is always there. In recent times some have argued for a "convergence" of science and religion, or to define them as mortal enemies, destined to be in a "conflict" that brooks no discussion. It is true that some overlap will always take place, as seen in a great scientist such as Isaac Newton, who mixed theology with his science. There is no denying that science is a human enterprise, and its practitioners—whether believers or atheists—bring personal motivations to their work. However, modern science rejects any theological intrusion into its theories or experiments. There are different levels of human knowledge, often coming from common roots in human nature and history, but also having their autonomy.

This might be called a "complementary" approach to science and religion, which implies that each has territories with limits, much as human knowledge will have limits. The complementarity also recognizes that all forms of knowledge are based on metaphysics—metaphysics being the larger assumptions that any thinking human being makes about the very nature and essence of the universe. Even so, a complementarity approach calls for boundaries as well. It does not allow for a scientific religion or a scientific theology, or for religion to offer alternative explanations to scientific theories.

Once the metaphysics are recognized, and the proper boundaries between kinds of knowledge are respected, people of faith and of science can speak more openly, if they care to. The religious believer is free to speak and reflect from within a personal faith commitment, and in conversation with the scientist, to discover patterns and accomplishments that may turn out to be complementary and enhancing to a religious worldview. Religion may feel free to hold that its spiritual metaphysics encompasses even science, while science may hold that its apprehension of facts in the universe is crucial to all the world faiths, whatever their culture. This view allows religion and science to be equal partners in a democratic, interdisciplinary conversation where science is taken absolutely seriously, but where the voice of authentic religious commitment is also clearly heard.

In the following nine essays, chosen from the Templeton Science and Religion Series of books, the reader is invited to share in this complementary discussion between science and religion.

We start with astronomer and cosmologist Joseph Silk, who tells the story of the big bang, how it became our standard model, and the human drama involved in that discovery. As Silk says, we will never have experiments as powerful as the big bang to operate on Earth, so understanding its event and consequences continues to be one of the greatest experimental tools. As we study the universe and its great age has been revealed, we have realized that if the initial conditions of the universe had been slightly different, there would have been no structures and no life. Still, what remains unknown are the universe's dominant components—dark matter and dark energy.

In the second selection, paleontologist Ian Tattersall takes us back to consider the formation of the ancient earth, and how it has provided us with a key to our knowledge of the past: fossils. The fossil trail is empirical and speculative, because scientists have had to organize the evidence over a few hundred years of research. The idea of a "tree of life" has continued to evolve, now obtaining a more complex structure than we had once thought (now like a bush or tangled vines). The scientific tools for dating fossils have been improving, and new fossils continued to be unearthed. The question of the first "life" still eludes us, however, even though scientists and a good deal of speculation have given some of the best possibilities for how life arose in an inhospitable environment. Life requires complexity, but there is no definitive understanding yet of how complex organisms evolved into living organisms.

Environmental geneticist R. J. Berry next tells us about Earth as a "green machine." The belief in the creation of the world for our benefit has been an obstacle to the study of the natural world and gave us reasons to abuse it. Scientific naturalists of the past, while keeping religious beliefs, began to study nature for its own integrity. This led to our ecological sciences. Along the way, Western culture had to reconcile a new view of the Earth with that of the Bible, which told a different story about the Earth's age and formative stages. Berry tells how this great transition took place, and about some of the historic characters, such as Charles Darwin, involved in the new science of "biogeography": how biological diversity has spread across the Earth, a key topic for ecology today. We have realized that diversity is driven by genetic variation.

Then in chapter 4 we turn to the science of human nature. After their long careers in psychology and neuroscience, Malcolm Jeeves and Warren Brown have much to tell us about how humans are like our closest animal relatives, and yet different by a "quantum leap." How did this happen, and how do we make sense of this, so that, as the authors say, we do not see humans as "nothing but" smart animals? The main distinction lies in the use of language, the authors say. They introduce topics as diverse as language formation, "theory of mind," and altruism. They see no problem for religious belief in studying us as a "human animal" as well. The authors argue for a more complex and holistic view of the brain and warn that increased complexity of the nervous system does not always mean increasing complexity of learning capacity or social behavior.

After that, geneticist Denis R. Alexander has a fascinating story to tell: how biological evolution could not make sense until we mastered the science of genetics. "Genetics is what rescued Darwinian natural selection from oblivion," he says. This is the story of the theory of natural selection looking for a source of "variation" in the characteristics of living things. The genetic pool was the answer, and from the monk Gregor Mendel to some of modern biology's most brilliant researchers, we finally arrived at the "modern synthesis." Alexander shows us how these two-steps of evolution work together in the field and in the laboratory. He also shows how the same concept—such as chance—can have completely opposite interpretations depending on political and cultural circumstances.

Cognitive scientist Justin Barrett then asks how we can make sense, from a scientific viewpoint, of the enduring nature of religion among human beings. It has its origin in the very way our brains and five senses work, he explains. This cognitive approach to religious belief and experience has social implications, shaping cultures and building theological systems and religious institutions. To explain this, Barrett takes us back to the basics of human "cognition" and perception. He shows, perhaps surprisingly, that religious belief is only "natural" for the way the human being is constructed. Humans have a tendency to look for intention or causal agents where "abstract causations" fail.

In chapter 7, mathematician and computer scientist Javier Leach introduces us to mathematics as a "language," one that complements our other languages (natural or scientific, for example). He explains how the very nature of mathematics has developed some of our theological concepts, such as infinity and God's existence, involving cognitive processes that give us the sense of absolute consistency. To illustrate, Leach explores the famous "ontological argument" for the existence of God, an argument made by Jewish, Christian, and Muslim thinkers based on a similar system of logic. Although the "proof" is not universally persuasive, Leach says, it shows the common importance of logic for both science and theology.

Noreen Herzfeld, a computer scientist and scholar of religion, introduces us to cutting-edge science that is both ethereal and very small. Computers are forcing us to view human intelligence in new ways, she says, and cyberspace may be altering our human perceptions of reality, even permanently, as well as our perception of religion: How can we conceive of divine action in a virtual space? Meanwhile, nanotechnology is probing how to make molecular-size machines, a scary idea for many people. No less than that, there is talk of transforming our physical selves to defy death. By looking at these many questions in both artificial intelligence and nanotechnology, she introduces us to the promise and peril, and especially the religious and ethical decisions we must make about how far to go with these new technologies.

We end with medical science. After years of experience as a doctor, Harold G. Koenig argues that science needs the clearest possible definitions of "religion" and "spirituality" to understand how these affect health and medical practice. The days when the spiritual side of patients was ignored are over, Koenig says, especially because we must have every available resource in the coming "health-care crisis." Koenig not only makes finer distinctions, such as dealing with patients in their own faith tradition, but provides practical tools for clinicians and patients. He shows that empirical research is on the side of taking religion and spirituality seriously in the world of medicine. This has become more urgent as troubled economic times require communities to play a major role in caring for patients.

All of the authors suggest that we live in a world of interconnectivity. This is a classic idea that goes back to the older mystical traditions, yet it seems to be borne out once again in the empirical sciences of today. Every single thing is connected to its immediate environment and to the rest of the world. We can identify singular things—a quantum, a star in a galaxy, a bee, or the seat of human intelligence, which some may call the soul. But none of these has meaning (or the ability to prosper) as isolated entities. Each acquires significance only in relation to other entities, either of its own kind or in the wider global environment.

A collection of essays such as this hopes to open the doors of research related to science and religion a bit further. The nine authors suggest some of plausible ways to draw boundaries in science and religion, and yet cross over. Scholars have their appropriate ways to do this, but the general public is also encouraged to look for the complementary relationships in their lives. For the scholarly world, the horizon for the science-religion discourse seems to be the continuing problems of epistemological and philosophical categories.

In both science and religion, there must be an acceptance that we often face the unknown, yet such an acknowledgment does not undermine our confidence to move forward. Science and religion are helpful prods to each other in this quest. Religion invites science to move out of reductionism, while science invites religion to recognize more precisely the world of empirical evidence. Religion can ask science how it will serve humanity and "do no harm" to nature, while science can present to religion tools, mental as well as technological, to make these beneficent goals possible. An immense opportunity is open to develop a new holistic and interdisciplinary intellectual era. We may be at the dawn of a new paradigm that cannot be claimed by either science or religion by itself, but is only possible because of their willing, good-faith interdisciplinary encounters.

Joseph Silk
Cosmic Origins

"THE UNIVERSE IS a wondrous place," Silk says in his book *Horizons of Cosmology*. It is a wonder also that we now have a fairly simple explanation of the universe's origin, elements, galaxies, and stars. We have found ways to measure these, making cosmology a practice of theory tested by observation. To observe the largest scale, science has produced ground and space telescopes; to see the smallest, it operates supercolliders. Our best laboratory still is the big bang itself, after which the characteristics of energy and matter developed into the world we see today. Given the limits of measurement in cosmology, its greatest thinkers have been inclined to speculation as well: ideas that range from multiverses to metaphysics. "Cosmology leads inevitably into considerations that have philosophical and even theological overtones," Silk tells us. Cosmologists are all too human as well. Even they ask: Why do we exist in this particular universe? Is this a coincidence? Do we influence our universe as its observers? For as Silk notes, "After all, if we did not exist, who cares?" The great mystery still before cosmology is the nature of dark matter and dark energy, which seem to make up 90 percent of the universe. To move ahead, speculative belief can never substitute for good science, and in cosmology, that begins with the "standard model" of the big bang.

[handwritten marginal notes:]

Evidence
1 • Red shift 1929
• Expansion of space
Einstein 1915
2 • Production of H_2, He_2
3 • Background microwave
radiation 1964,
2.73°C black body
4 • fluctuations →
galaxies etc
• inflation, high energy
physics, etc.

CHAPTER 1
Case for the Big Bang
Joseph Silk

THE IDEA OF an expanding universe was a shock to early astronomers, but now the jury is in: the universe is indeed expanding. This is the inevitable consequence of what the America astronomer Edwin Hubble observed (in 1929) as the "redshift" phenomenon. He saw that the light from distant objects in the universe shifted to the red side of the spectrum, which, according to the laws of light waves, means that objects are moving away from the observer.

Curiously, Hubble himself never accepted the radically new idea of an expanding universe, even though it stemmed directly from his work. He rather chose to accept galaxy redshifts as an observable phenomenon without any commitment as to their origin in terms of the properties of space. Perhaps he was confused by the models of other leading cosmologists, who were proposing a static universe. Here, it was suggested that a hypothetical field produced the observed redshift, and indeed in these static models, the universe was seen as devoid of matter, let alone expanding.

The systematic recession of the galaxies is now explained as being due to the expansion of space. Albert Einstein's theory of gravitation, which in 1915 spoke of a curved time-space that could either collapse or expand, certainly predicts this phenomenon. But rather than collapsing, why is space expanding?

This question takes us back to the initial conditions of an infinitesimal patch of matter from which the universe began. That matter

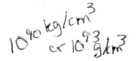

must have been in a volatile state, that is, out of equilibrium. This could have been a state of contraction or of expansion. Either way, the density of this primordial patch must have been ~~1090~~ grams per cubic centimeter. This is known as the Planck density, after German physicist Max Planck. This density is so high that it takes place only at the interface of quantum theory (in tiny atoms) and general relativity (large-scale gravity). In other words, at the initial conditions, the smallest and largest forces in the universe known today were squeezed together, united and indistinguishable.

The quantum processes were operating in the patch. By quantum jumps, macroscopic clumps of matter could disappear or reappear like the Cheshire Cat in *Alice in Wonderland*. Black holes, which are so dense with gravity that they attract all the matter around them, could have formed and decayed spontaneously. In this early state, the universe must have been at the most extreme density that can be conceived under known physics. It represents our best guess at the conditions that prevailed near the beginning of time.

After that, the direction of the universe has been quite predictable. It has expanded according to our basic measuring tool, the Hubble diagram, which plots distance compared to the velocity of galaxies as they move away from the central starting point of the universe. We deduce that this expansion began 13.7 billion years ago. The latest data, using supernovae to chart the expansion, have added something surprisingly new to the traditional Hubble diagram: the remotest galaxies are accelerating in recession, speeding up the expansion of the universe, a topic we discuss later.

The ancient age of the universe has also been a surprise to modern science, at least for a century or so. Today, scientists subscribe to the view of a very old universe of about 14 billion years. It is a difficult idea for a substantial minority of the population, especially in North America. Many people prefer a traditional interpretation of the universe drawn from a literal reading of the Bible. In one famous calculation from the King James Bible by seventeenth-century Anglican bishop James Ussher, the universe was created in 4004 BC on Sunday, October 23, at about 7:30 a.m. Today, decades of Gallup polls show that up to 50 percent of Americans think that human life arose fairly recently, according to a literal reading of Genesis, and for many, this would also include the belief in a very young universe.

Fortunately, from the time of Pope Pius XII in the 1940s, guided by the advice of astronomers such as Abbé Lemaître, the Catholic Church and other religious circles have taken a more enlightened approach to modern cosmology, which tries to find a proper balance between theology and science. This view holds that while science is paramount, it presents no challenge to a creed that rests on beliefs that arise from faith. Indeed the converse also applies: the beauty of science and the revelations produced by scientific discovery constitute part of the modern theologian's perspective and toolbox.

Today, for example, the discoveries of modern physics, astronomy, and cosmology reveal intricate details in the physical structure of the universe that seem highly improbable. The proton mass is remarkably close to the neutron mass. Were it very different, stars would not have formed. Further, the force that is accelerating the universe is far weaker than physics leads us to expect. Were this force much stronger, galaxies would never have formed. And in a universe devoid of stars and galaxies, there would not be any observers to marvel at the mysteries of the cosmos. It is not hard to see how theologians might find such discoveries fascinating.

These apparent coincidences in the universe have prompted some to argue that the arrival of human beings on Earth is perhaps not a cosmic accident after all. Indeed, those who employ this reasoning have elevated this human-centered argument into a fundamental principle that governs the universe, which has now been called the anthropic principle, for *anthropos*, or man. This principle has long held sway in traditional religion. But sadly, in the view of some, the wheel has turned full circle and now physicists too are appealing to the anthropic principle to account for the initial conditions of the big bang. Obviously, the anthropic approach is an unabashedly self-based egocentric worldview.

FOLLOWING THE EVIDENCE

Our concern now is the evidence for the big bang theory of the universe, for we do not want to take it just on hearsay. Four major predictions of the big bang theory have been verified by modern scientific experiments: the recession of galaxies, the abundance of light elements in the universe, the existence of a cosmic background radiation (blackbody) that is uniform, and finally, predicted rates of fluctuations in that same radiation. These four lines of evidence ought to be enough to quench even the most biased critics of what at first sight is a highly implausible theory.

Once the expanding universe had been predicted based on Einstein's theory of gravity, Hubble in 1929 measured this celestial movement. Less well known is Lemaitre's intervention in 1927 when he predicted and derived the very same law that Hubble, unaware of Lemaitre's work, had published two years later. Hubble went on to pioneer and greatly refine the correlation between recession velocities and the distances to remote galaxies. We judge the velocities by the spectra of light in galaxies. The redshift of spectral lines shows the recession velocity. Distances are measured by using so-called standard candles, objects (typically a type of star) that give off the same brightness and whose distances can be judged first in nearby cases, and then can be traced to the same kinds of stars at great distances. Hubble's successor at Mount Wilson, Alan Sandage, delved deeper into the universe with the aid of the two-hundred-inch telescope on Mount Palomar in southern California. By applying Hubble's law farther and farther out, Sandage realized that distant galaxies had recession velocities of 10 percent or more of the speed of light.

The second proof of the big bang comes from the prediction that the early explosion of the universe would produce light elements with the simplest atomic structure. That would mean mostly hydrogen, but also helium and traces of deuterium and lithium. In fact, the universe indeed is abundant with these light elements. This prediction came largely from the insights of George Gamow, a Russian refugee to the United States in 1934. Gamow was a nuclear physicist with a remarkably broad perspective. He initiated our modern understanding of thermonuclear fusion. How could a pair of protons merge together, he asked, to eventually form helium? Resolving this paradox led to our understanding of how stars shine and to the development of the hydrogen bomb.

Eventually Gamow became interested in the big bang theory. Until then, the prevailing view was that the big bang was a cold event that, nevertheless, led to expansion. But the cold theory had a problem. At the cold state, which means a very high density, atoms merge into the most tightly bound nucleus, which is iron. Hence, iron should be the most plentiful atom in the universe, which is certainly not the case.

Gamow realized that a moment of extreme heat could circumvent this problem. The first nuclear reactions, which produce new elements, could not begin until the nuclei of atoms had overcome the heat and then begun to merge with others. At the start of the universe, he reasoned, there was only a small space of time, perhaps only a few minutes, for the light elements so abundant in the universe to form before the cooling down produced the heavy elements. At some early instant, the universe had produced hydrogen, which today remains the predominant element. It must have been a hot beginning.

Following this logic, Gamow was the first to see that cosmology presented the ideal conditions to understand the origin of light elements. Of course, for the light elements to have seeded the many other nuclear reactions that followed, the universe must have been exceedingly hot. But no one had detected any residual heat in the universe, as predicted by a hot big bang, so the prevailing wisdom favored a cold origin. Decades later, the cold theory was still being advanced by the great Russian cosmologist Yaakov Berisovich Zel'dovich, and Gamow's contribution was mostly forgotten.

However, lack of evidence for a hot universe did not deter Gamow's genius for advocating a new way of doing cosmology. He now turned to stars. Gamow believed, erroneously as it turned out, that stars did not have the thermonuclear power necessary to synthesize the chemical elements seen in the universe. He was partly right. Indeed, stars cannot make a significant quantity of light elements, although they can make some. With stars eliminated, Gamow argued, the universe itself was the ideal place for synthesizing the second most abundant element, helium. He proved his prediction by describing how the universe expanded in phases, briefly achieving temperatures above those in the Sun. A phase needed to last only a few minutes, he argued, to produce light elements.

This takes us back to the start of the universe. The only massive particles that could have lived on from the beginning were protons, neutrons, and electrons. Protons and electrons make up hydrogen, which is why that is the most abundant element today. Initially it was too hot for atoms to exist. The dominant constituent of the universe was ionized hydrogen, which meant it now had an electrical charge and was the only chemical element in the early universe. A few neutrons, about one for every proton, were present. For the universe to move toward formation of other chemical elements, a great deal of heat was needed so that nuclear reactions could combine protons and neutrons. With the heat, protons could overcome the force of Coulomb repulsion (that keeps particles apart) and form elements heavier than hydrogen. Gamow had discovered this natural barrier. With his insights, we find the beginning of nuclear physics as a new branch of science. His brilliant idea was that if the universe were, by fiat, initially hot, the required nuclear reactions would have taken place in the first minutes.

He enlisted his student Ralph Alpher and his colleague Robert Herman into the research project, which culminated in predictions of the exact abundance of light elements in the universe. Helium, amounting to some 30 percent of the mass in the universe, was synthesized in the first minutes of the universe. Previously a mystery, the origin of helium—the second most abundant element—was now resolved.

Gamow's dream of the origin of all of the chemical elements had one problem, however. Only 2 percent of the elements produced in his hot big bang were heavier elements, such as traces of lithium and beryllium. Where did all of the heavy elements come from? As we now know, the heavier elements are made in exploding stars called supernovae, which scatter their ashes around the galaxy. Nevertheless, Gamow was fond of joking that his theory should be considered a success, and rightly so. It explained the nature of 98 percent of the matter in the universe: hydrogen and helium. However, the most important proposal of his theory, the expectation of a hot universe, was to remain forgotten for nearly two decades.

That changed in 1964, when the background radiation of a hot big bang, once predicted, was now discovered quite by accident, resulting in our third piece of evidence for the big bang. In New Jersey, the radio astronomers Arno Penzias and Robert Wilson had gained access to a microwave radio telescope. It was originally designed for the first satellite communications system, but was subsequently overtaken by better technology. Using this old equipment, Penzias and Wilson wanted to survey only the Milky Way galaxy. But they found beyond the Milky Way a pervasive glow that is apparently isotropic, that is, has the same structure in all directions. (Being isotropic is not necessarily the same as being homogeneous, however.) Furthermore, the glow seemed to have no relation to our galaxy. They at first refused to believe their measurements. They tried to explain it away as an experimental artifact but they did not succeed.

Then a bit of news put them into action. When they heard that a rival group at Princeton was searching for the fossil glow of the big bang, Penzias and Wilson realized what they had discovered. They promptly published their measurement of excess radiation—the beginning of a long process of proving Gamow's hot big bang theory. Indeed, Penzias and Wilson were initially unaware of his arguments. Nevertheless, they had found the elusive background from the beginning of the universe.

At last, Gamow and his collaborators were vindicated. However, they never received full recognition for their prediction of an initially hot universe that produced relic radiation. They did not really appreciate the need to connect their theory with microwave astronomy. In their publications as well, they did not use terminology that would have caught the attention of microwave astronomers, such as Penzias and Wilson. I recall once encountering George Gamow surrounded by a small crowd of astronomers, as he declaimed in his high-pitched voice that he "had lost a penny, Penzias and Wilson had found a penny, and was it his penny?"

After the serendipitous discovery by Penzias and Wilson, much of the early debate about the hot big bang came to a climax in early 1967. The chief forum turned out to be a cosmology conference at the Goddard Institute for Space Studies in New York, where the topic was intensely debated. For historical reasons, this meeting was called the Third Texas Symposium on Relativistic Astrophysics, following earlier meetings in a series at Dallas and Austin. These were heady days in astronomy and cosmology. At the first Texas meeting in 1963, the superstars were quasars, just discovered (although their true nature and distance still are debated). The newly named idea of a black hole, a singularity in space of nearly absolute density, much like the start of the universe, was also announced at one of the Texas symposia.

The 1967 New York meeting marked a turning point for acceptance of the big bang theory. Before then, the very name was a kind of slur. It was coined pejoratively by British cosmologist Fred Hoyle, who favored the rival steady state model of the universe. As Hoyle famously told a popular BBC radio broadcast in 1950, the idea of "big bang" was "an irrational process that cannot be described in scientific terms . . . [or] challenged by an appeal to observation."

Even after the discovery made by Penzias and Wilson, however, it took another fifteen years to verify the exact nature of this cosmic background radiation. According to predictions, it had to be what we call *blackbody radiation*. A corollary of the light element interpretation was the prediction that the cosmic microwave background would show a blackbody spectrum, which has so completely mixed the different wavelengths of heat that it approximates the evenness of a perfect furnace. The first prediction of this perfect blackbody temperature was made by Alpher and Herman. But applying such calculations to microwave astronomy did not come for another two decades, when Robert Dicke at Princeton (the rival whom Penzias and Wilson had heard about) was arriving at the best estimates of the background radiation temperature.

The blackbody spectrum would be expected if it emerged from a dense and hot beginning. The principal attribute of a blackbody spectrum is that it carries no information about the sources of heat and radiation. Everything is mixed, where just as in the ultimate car wreck, a Ferrari cannot be distinguished from a Ford. This is a state of maximum entropy, or disorder, and hence we have the absolute minimum of information. According to the second law of thermodynamics, a fundamental law of physics, the order of a dynamic system, left on its own, can only remain constant or decrease. Chaos reigns. But as structure develops, the entropy is decreased. Curiously, as stars and planets form, creating structure, they also radiate heat, which is like expelling the entropy from the system. This is exactly what happened in the formation of the Milky Way galaxy, for example. The entropy of the universe as a whole is conserved, but at the same time, we see the rise of structures of increasing complexity, such as galaxies, stars, planets, and life itself.

The entropy of the universe is measured by its radiation content. We know this because the number of photons in the cosmic microwave background vastly outnumbers the number of particles in the universe. This is true for any high-entropy system. Blackbody photons are the ultimate cosmic equalizer in terms of lack of any preferred information content.

The first definitive spectral measurement of the microwave background was made by the Cosmic Background Explorer (COBE) satellite. The satellite carried sensitive experiments that were designed to measure the frequency distribution of the sky photons as well as tiny differences in the sky temperature to a level better than anything previously accomplished. This accomplishment did not come quickly. It took seventeen years to launch the satellite after it was first conceived. Long delays are common in space astronomy, but this one was exceptional. The delay of COBE was compounded by the *Challenger* space shuttle disaster in 1986. With the curtailment of space shuttle missions, NASA had to reconfigure the COBE satellite for launch on a Delta rocket, which took place finally in 1989.

One experiment on the COBE satellite measured the spectral energy distribution of the cosmic background photons, which is a clue to their origin. The Penzias and Wilson measurements suggested it would lie almost entirely in the far infrared and microwave frequency region. This very red distribution corresponds to extremely cold radiation, at only 3 degrees Kelvin. Compared to early instruments, however, COBE's measurement of the spectrum was remarkable for its precision.[1] The temperature was measured to be 2.736 degrees Kelvin. It is a perfect blackbody. In fact, a better blackbody spectrum cannot be synthesized on Earth. The universe emerged from an ideal furnace.

A few seconds after the start of the big bang, the temperature was about a billion degrees, but the universe began to cool quickly. The blackbody spectrum found by COBE corresponds to what the big bang would have been like for the first few weeks, a perfect furnace. During this phase, the universe was completely opaque. Matter was in the form of hot plasma. The plasma was mostly ionized hydrogen, produced when the electrons cooled enough to be captured

by protons. Soon hydrogen atoms became the dominant constituent of the universe, at least in terms of ordinary matter. Hydrogen
is ineffective at scattering the radiation, and so about three hundred
thousand years after the big bang, the universe became transparent
(at a cool temperature of about 3,000 degrees Kelvin). We can see
freely back to this point in time. As we measure the radiation today,
we are seeing back to that early phase of the universe, just by looking in a dark patch of the sky.

The Final Piece

We now come to the final line of evidence that has been crucial in
verifying the big bang: confirmation of the fluctuations in the cosmic microwave background. Without these, the entire edifice of
the big bang might have collapsed like a house of cards. We needed
to explain the original seeds of the structure we see in the universe
today, and after a long search these seeds or fluctuations have been
found.

The fluctuations must have arisen at an early phase of the universe, when matter no longer was in competition (by friction)
with the radiation. The matter would naturally be slightly denser
in some regions, compared to the average background, and thus
have the gravity to attract more matter. This slight overdensity
increases. This movement is not in one direction, however, because
every overdensity is matched by an underdensity. The underdense
regions become emptier. The overdensities are the precursors of
galaxies. This early back-and-forth of density produced a pattern of
fluctuations that should, as astronomers predicted, show up in the
background radiation.

The physics of density fluctuations has proved to be remarkably
fruitful in all areas of cosmology, from the beginning of the universe
to its large-scale structure and the formation of galaxies. It all begins
with the very nature of gravity. Gravitation is intrinsically attrac-
tive. Pressure opposes gravity, but in a large enough system, gravity

is overwhelming. It results in fragmentation. We refer to this process as *gravitational instability*. Overdensities grow at the expense of
underdense regions. By analogy with capitalism, the rich get richer.
When we look at the universe today, we see structures, which tells
us that there must have been gravitational instability at the very
start of the expanding universe. In other words, the universe could
not have been completely uniform when the expansion began. If
it were uniform, structures such as our Milky Way galaxy would
never have developed. Finite density fluctuations must have been
present from the beginning in order for structure to have evolved.
But the predicted value of the seed fluctuations was a fraction of a
percent in the very early universe.

Once the cosmic microwave background was discovered, beginning in 1964, cosmologists finally had a backdrop against which
to measure accurately the fluctuations of the early universe. These
infinitesimal variations are the only way we have to probe the initial conditions of the universe. In principle at least, the fluctuations
would show up in tiny differences in temperature. The fluctuations,
as ancient relics, would reflect the kind of large-scale structure and
dispersion of galaxies we see in the universe today. In other words,
today we see the relics of the big bang in only three forms: one
being the light elements, the next and complementary form being
the cosmic microwave background, and the third being the structure of matter. We call the heavy particles that produced the first
matter *baryons*, a general term for the protons, neutrons, and electrons that make up every element.[2] Baryons are relics of the early
universe. They are the stuff that constitutes our bodies today.

Once the idea of fluctuations was clear in cosmology, the race
began to find them in the sky. Over the past decade, these measurements have become increasingly accurate. When it comes to predictions on the kind of fluctuations that seeded the early universe,
in fact, no significant inconsistencies have been found. In all, this
has been a final dramatic verification of the big bang theory.

The research on fluctuations has given us a great insight into the formation of galaxies. In confirmation of the big bang, we have seen that the distribution of galaxies shows us what the fluctuations must have been like back at the beginning. We observe a highly uneven distribution of galaxies, so we can assume that such uneven irregularities must have been present before the galaxies formed. Now we move forward and imagine how this process has taken place.

Imagine a region that is large enough to contain all of the mass that now makes up a galaxy, but is very slightly overdense, by a fraction of a percent. In fact, this seems small but is nevertheless highly improbable if left to chance. As noted earlier, density fluctuations are able to grow in strength. The slightest excess in gravity attracts surrounding matter. A small overdensity eventually becomes larger. This region of density still expands along with the universe, but at a slower rate, so it lags behind. When the contrast with its surroundings is large enough, the region becomes dominated by its own self-gravity. This region is cooling down, which means its outward pressure is being overcome by the gravity of its matter. At this point, the region is gravitationally unstable. The region must collapse to form a massive cloud, which becomes the birthplace of future galaxies.

Inside the cloud, the density fluctuations also cause it to begin breaking up into clumps. Even in a static cloud, infinitesimally small fluctuations could grow exponentially. A slight excess in density would attract matter from its vicinity. The fluctuation gains strength. But now, the expansion of the universe means that any region about to collapse into a cloud is itself expanding. It expands until its own self-gravity becomes dominant. The expansion quenches the rapid growth, but slow growth still occurs. Overdense regions continue to become denser by drawing matter from the surroundings, but more gradually than in the case of a static cloud.

The expansion guarantees that it takes a long time for the overdense region to form a self-gravitating cloud. It is as though one is running on a continuously expanding track. It takes much longer to finish the race. Nevertheless, at the start of the universe, such clouds are destined to form the first galaxies. The typical masses of galaxies are initially small, only a few tens of millions of solar masses. In time, however, larger and larger clouds condense, and after a billion years most galaxies are in place. The universe is then one-fifth of its present size.

Galaxies continued to form over the next 10 billion years. As mentioned earlier, we can tell the younger and older galaxies by their shapes and colors. The older galaxies have elliptical shapes with predominantly older stars. The younger galaxies have spiral structures. They are disk-shaped and often have a central spheroidal bulge of old stars at the center. Galaxy colors also tell us the ages, and colors depend on the mix of stars. A blue star is massive, hot, and short-lived. A red star has low mass, is cool, and is long-lived. We find that spiral galaxies are blue whereas elliptical galaxies are red. In fact, the old red stars are the dominant constituent of most galaxies, having formed when the universe was young, and also forming the spheroidal component at the center of galaxies. Disks are intrinsically blue, although partly reddened by interstellar dust and by the underlying component of older stars.

Today, star formation continues in the gas-rich disks of galaxies, and especially in the spiral arms. Ongoing star formation traces out beautiful spiral patterns. Our Sun, for example, is at the edge of the Orion spiral arm in the Milky Way galaxy. The nearby Orion nebula demarcates the most active star-forming region in our neighborhood. It is a nursery for star formation.

We have come a long way: from the big bang furnace through the early fluctuations to the large-scale structure of the universe. The one thing that was not anticipated, or even understood until very recently, is that the expansion of the universe is accelerating. According to Einstein's theory of general relativity (that is, his theory of gravity), gravity should be causing a deceleration of the expansion of the universe. This is a positive pressure that enhances gravity, moving the universe toward collapse. But we find that the universe is not only expanding but it is also speeding up. Something is providing the negative pressure that counters gravity, causing the acceleration and leading to one of our most baffling theories in modern cosmology—a negative pressure, or cosmological constant, that we call dark energy for lack of a better term.

Overall, we have just covered what is now the standard model of cosmology. It brings us to one more remarkable conclusion about the universe. It has reached the critical density that is required for the universe to be structured according to Euclidean geometry, meaning we can measure the universe as if it were flat. But meanwhile, much of this flat universe is made up of something we cannot see, namely dark energy and dark matter. Two-thirds of the mass-energy density of the universe is made up of dark energy, according to our calculations. We don't know what dark matter is made of, but we know that it makes up the other third of the density of the universe. The mystery only deepens, for just 15 percent of the dark matter is composed of baryons, the stuff that forms stars and planets. Cosmology is deeply challenged by the quest to decipher the exact nature of dark matter, or to prove whether it even truly exists.

Particle Physics and Inflation

Up to now, the great theorists of our universe have been astronomers and cosmologists. They are immortalized for various models of the expanding universe, at least before we arrived at the conventional model of today. What they had always lacked, however, was a theory about the initial conditions of the universe, in the veritable first moments of the hot big bang. The cosmology coming out of the 1930s did not have a theory. A new idea was needed. It finally came from a group of scientists who hailed from outside of astronomy, the particle physicists.

As outsiders, their arrival jarred astronomy but also was a new impetus for advances in cosmology. Particle physics was the field of nuclear physics that pioneered the understanding of the origins of helium, for example. The particle physicists looked at the big bang as a great particle accelerator that had achieved the highest energies possible in the universe. It had conducted one great experiment long ago, and we can now look at the consequences. To do particle physics, they had to think this way. On Earth, it is hard to imagine constructing any particle accelerator more than a hundred kilometers across. The size of such technology is limited, and so are budgets. For a machine to match the energies at the first instants of the big bang, for example, we would need an array of superconducting magnets that stretch to the moon or beyond. The benefits of focusing on the beginning of the universe became obvious. One can study the effects of ultrahigh energy particle interactions for free, or, at least, for a relatively trivial sum.

This led to the field of inflationary cosmology, which first emerged in the 1980s by asking the question: Why is the universe, despite its likely irregular structure at the beginning, now overall fairly smooth, even, and flat in the Euclidian sense? The answer seemed to be that at the start of the universe, there might have been a rapid inflation of matter, evenly mixing up its forces and particles before it began its path of expansion.

How did they arrive at an inflationary theory? It starts with basic physics. At very high energies, some new physics comes into play. High-energy physics uses the analogy of symmetry. All fundamental forces have the same strength. However, as the universe cools and expands, this symmetry is broken. We can imagine symmetry by thinking what would happen to a Rolls-Royce and Volkswagen as each is exposed to extreme heat. At first they remain distinctly different. Soon, they would melt into a soup of inorganic chemicals. No doubt there would still be differences. But at even higher heat, one would be left with the constituent chemical elements. The cars would be indistinguishable. Symmetry prevails. This was the beginning of the universe, paradise for some but closer to hell for others.

We can also go in the other direction, which is the breaking of symmetry. This is the story of the formation of galaxies. As the temperature of the universe drops, chemical and physical forces separate because the force that holds the atomic nuclei together is stronger, and on a smaller scale, than the force that holds atoms together. The cooling also allows the electrons that are now orbiting the nucleus, like planets around our Sun, to begin a process of electromagnetic interactions, which control chemistry. This allows hydrogen to form. The cooling universe is on its way to forming stars.

All of this has happened because the original symmetry between the fundamental forces of the universe broke down. Let's start at the very beginning of that process. At the highest energies there were no nuclei, not even protons or neutrons. The natural state of matter was a soup of quarks and gluons. A proton is made up of quarks that are held (or "glued") together by clouds of gluons. We are in the realm of quark-gluon plasma, the dominant state of the universe about a tenth of a nanosecond after the big bang. The phase transitions continue, each a change in the state of matter that is accompanied by a release of energy. It is much like the energy given off when ice melts into water, a global warming if you will.

The next instant is what we call the era of electroweak unification, when the electromagnetic and weak nuclear forces of the universe were in total unity. This moment comes an instant after the big bang, a fraction of time so small that it has fifteen zeros after the decimal point: 0.0000000000000001 of a second (one-quadrillionth of a second). This was not a complete unification, however, for the strong nuclear force had already broken away from the party. We call this earlier moment the epoch of grand unification when electromagnetic, weak and strong, nuclear forces were all identical. Gravity was united with the other forces even earlier, but that was at a still higher temperature, which we call the Planck instant, also after German physicist Max Planck. This realm of Planck physics is poorly understood, so for the moment we bypass this very beginning point of symmetry.

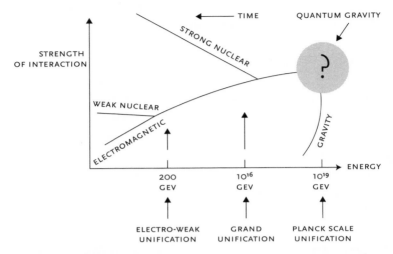

FIGURE 1. The breaking of symmetries from quantum gravity, to grand unification, and to the appearance of our low-energy universe when the electromagnetic and nuclear forces separate.

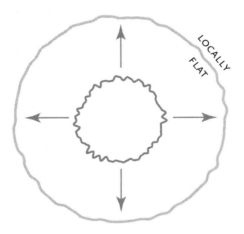

FIGURE 2. How inflation removes all initial irregularities in the universe and simultaneously explains why it is so large and spatially flat.

Returning now to the moment of grand unification, it too breaks down as the temperature drops. More symmetry is lost. As mentioned earlier, the strong nuclear force becomes stronger than the weak force. They break apart. The weak force remains indistinguishable from the electromagnetic force until much more cooling occurs. Then the weak force separates. The electromagnetic force is also free to operate, based on the electrons that orbit around a nucleus of the atom. Chemistry has begun. Its first product, much later, is hydrogen. So many phase transitions are happening at this point in time that a very substantial amount of energy is injected into the universe. One of these releases was so powerful that it inflated the size of the universe in an instant. The heart of inflationary theory is that the inflationary release of energy does not last long, but it manages to provide a big transient boost to the energy content of the universe. The energy release counters the effect of gravity in a dramatic way. The effect produces repulsive gravitational forces, which act like antigravity. The expansion rate of the universe is boosted. There is so much antigravity that exponential expansion sets in. Hence, we call this *inflation*—indeed, the mother of all inflations.

We can imagine this transformation taking place in an infinitesimal patch of the universe. It can be anywhere, but it seems inevitable somewhere. Once such a patch develops, it switches from having a strong gravitational attraction to having a strong gravitational repulsion. Powered by this phase transition energy, the patch grows exponentially, inflating the universe just as suddenly. In fact, this infinitesimal patch has now encompassed almost the entire universe. By virtue of this rapid and immense expansion, the universe becomes almost completely uniform.

This answers the baffling mystery of the universe's general uniformity. With the inflation there is an associated flattening of the geometry. The geometry of space may once have been non-Euclidean. Perhaps all space was curved before the moment of inflation. But after an immense expansion, space flattens out, at least locally. This is what we mean by an ironing out of possible wrinkles in space-time. Space is wrinkle-free, and it is much, much larger than the observable universe today. Inflation gives us a reason. The universe may once have been a messy environment. It could have expanded more rapidly in one direction, leading to a cigarlike shape of space. Or in two directions, resulting in a giant pancake. However, inflation smoothed away nearly all such deviations from uniformity and sphericity.

All of this occurred very soon after the big bang. The way we measure this event is by thinking of about a trillionth of a trillionth of a trillionth of a second—a decimal point followed by thirty-five zeros, or $.10^{-36}$ seconds. Let's call this a googolth of a second.[3] Once ten googolths of a second have elapsed, inflation is over. The phase transition is complete and inflation stops. The universe is now very large but very cold. Fortunately lots of energy is left over from the inflationary epoch, which causes the universe to reheat, and the usual hot big bang resumes as the universe continues to expand.

We have gone back nearly to the beginning, but there is a barrier in time that we cannot penetrate. This is even earlier, about a ten-millionth of a googolth of a second after the big bang, the sought-after instant when gravity was joined with all the other forces. At this moment, the strength of the gravitational force equaled that of the nuclear forces. This is the Planck instant.[4]

We have no theory that describes this state, which is where quantum theory meets gravity. Arriving at the theory of quantum gravity, the ultimate unity of forces, is the ultimate goal of modern physics. Perhaps this theory will provide the explanation of why inflation occurred. Right now, inflation is a very good conjecture, one that has predictable and verifiable consequences. For us, the most important of these is the emergence of structure.

In 1980, particle physicists such as Andrei Linde, then in Moscow, and Alan Guth at the Massachusetts Institute of Technology provided us with some of our most innovative theories of how inflation took place. They were outsiders to astronomy, but they too realized that the big bang provided a natural particle accelerator. It had achieved energies vastly higher than any man-built machine could ever accomplish. For example, the energy at the big bang was 10^{16} GeV. We can contrast this with the largest terrestrial machine we have, the Large Hadron Collider, a twenty-six-kilometer (seventeen-mile) circumference particle accelerator straddling the Franco-Swiss frontier. It began operation in 2009 under the supervision of the European Organization of Nuclear Research (CERN). As big as it is, however, the accelerator generates energies of just a few GeV, less than the big bang by a factor of more than a thousand trillion.

The Large Hadron Collider is likely to move us closer to an understanding of the early universe. But as a result of the energy discrepancy, our prospects of verifying grand unification and the phase transition that triggers inflation are remote. In the meantime, the best we can hope to do is to pin down the outcome of inflation. At least two distinct signatures can help us distinguish between competing models of inflation. Both come from the inflationary epoch.

One is a background of gravity waves. These are produced at the end of inflation along with the density fluctuations that seed large-scale structure. The difference is that gravity waves do not seed anything. No compression is involved as a gravity wave passes by. The wave only has a shearing motion. Still, these waves are a signature of inflation. The difficulty in detecting them is that they are highly redshifted. The wave frequency observable today is a thousandth of a hertz or less. Terrestrial backgrounds interfere at such low frequencies. These include the ocean tides. To detect such low frequencies, one has to go into space and use satellites that are millions of kilometers apart. Laser beams connect the satellites. As a gravity wave passes, one has to measure a change in the length of the laser beam of a centimeter or less to arrive at the expected frequency range where a signal might be detectable. Such an experiment is under study by the National Aeronautics and Space

Administration (NASA) and the European Space Agency (ESA) for a possible launch in the next decade.

Another signature of the original inflation is the distribution of fluctuations in the cosmic microwave background on angular scales more than a degree. New experiments will study these fluctuations and look for the inflationary imprint of gravitational waves on the fluctuations. One effect is a slight polarization. To detect this effect, a sensitivity of a hundredth of a millionth of a degree Kelvin is required, a hundred or more times better than our currently achievable sensitivities. Future space experiments will be needed to achieve this goal.

The search for inflation has taken us into space. It has also forced us to try to explain the large-scale structures of the universe.

Notes

1. This was the Far Infrared Infrared Spectrometer, designed by a team led by John Mather, who shared the 2006 Nobel Prize in physics with George Smoot.
2. Light elements are also baryons, in particular helium, which constitutes one-third of all the baryonic matter.
3. Actually a googol is 1 followed by a hundred zeros. Google founders Larry Page and Sergey Brin based their company's name on "googol."
4. The Planck instant is 10^{-43} seconds.

Ian Tattersall
The Ancient Earth

IN HIS BOOK *Paleontology*, Tattersall reveals the challenges of tracing the past through fossil remains. Yet the picture we have is one of a consistent and connected "tree of life" over 3.8 billion years. "The fossil record is vast and constantly growing," Tattersall says of ongoing research. Even though evolution of life takes place in the moment, with no plan for the future, in hindsight the fossil record can give the impression that life forms "progress." What is certain is that, among all species, humans are the ones able to ponder the world's immense biota and remarkable antiquity. We ask how it all began, an event that "lies beyond the bounds of today's science," as Tattersall says. In the meantime, we produce philosophical and religious answers. When that happens, says Tattersall, some scientists like a "good fences makes good neighbors" policy between science and religion. Other scientists see science and religion as two aspects of the same human curiosity—like a two-stage rocket: "Starting firmly in the material world, you can ride the scientific first stage to the point at which its fuel is exhausted, the point that lies at the limits of testable knowledge. From there—if you wish, or feel the need, as most people do—you can ignite the spiritual second stage, and be transported to the limits of the human ability to understand."

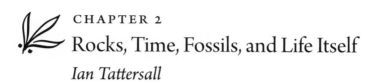

CHAPTER 2

Rocks, Time, Fossils, and Life Itself

Ian Tattersall

WHETHER LIVING FORMS exist elsewhere in the cosmos or not, for all practical purposes life as we know it was born here on Earth several billion years ago. An awful lot has happened since then, and it is in the rocks composing the surface of our planet that we find the fossils that document the long history of living things. So it seems appropriate to begin this chapter on paleontology, the science that studies those fossils, with a few words about the planet that we take so much for granted.

The geologist Preston Cloud once neatly described our Earth as an "Oasis in Space," which is, I think, about as apt a short description as it's possible to achieve. Our planet today really is an extraordinary place, with an oxygen-rich atmosphere, abundant water, a hospitable range of surface temperatures, and all the other necessities for the maintenance of life as it is familiar to us today. This amazing and comfortable environment exists, moreover, in the midst of a vast, hostile emptiness. Yet life itself came into being under very different—and very much more extreme—conditions.

The matter of origins goes back in an infinite recession, to a point that lies beyond the bounds of today's science. But scientists know the general outlines of how the Earth first began to form, some 4.5 billion years ago, out of a roiling cloud of hot dust and gases that eventually condensed to form our solar system. In early days the Earth's surface was an inferno, assailed from below by raging radioactive heat and from above by a constant bombardment of asteroids, as the remains of the debris cloud were "mopped up." Volcanoes on the hardening surface vigorously exhaled gases such as carbon dioxide, ammonia, and methane into an atmosphere initially consisting largely of hydrogen and helium. In brief, the early atmosphere was a toxic mixture of gases that would have been hostile to almost all forms of life that we know today. Equally inhospitable were the noxious oceans, which started to form as soon as the Earth's surface had cooled sufficiently to support liquid water, initially gassed out as vapor.

Still, the formation of the planet's solid outer crust proceeded rapidly as the fireball lost its initial heat. The earliest rocks known may be as much as 4.3 billion years old, and are believed to be witnesses to the early operation of the processes that have governed the form of the Earth's surface ever since. Once the crust had hardened sufficiently, its surface began to be cracked by the motion of the hot, molten rock below. Imagine a pot of thick soup simmering on a stove. Warm soup rises from the bottom of the bowl at the middle, where it is hottest. On reaching the surface it flows outward to the sides of the pot, where it cools and sinks once more, ultimately to be reheated and rise again. Driven by the radioactive furnace in the planet's interior, an identical process was established under our feet well over 4 billion years ago. The upshot is that the surface of the planet was, and continues to be, divided into a varying number of more or less rigid tectonic plates that are forever in motion. New, hot magma is added on one side of each plate as it is erupted along the linear structures known as midocean ridges, while old, cold rock is returned to the depths along subduction zones at the other side.

The basaltic rocks of the oceanic crust are relatively heavy. As a result the lighter rocks that compose the continents "float" above them and stand high above the ocean basins like giant icebergs. The floating continents are passively carried along on the "conveyor belts" below, like logs in a current. When one of them reaches the side of the plate on which it is sitting, it may bump into and crumple the continental mass on the adjacent plate. Forceful collisions of this kind have produced the great linear mountain chains of the world such as the Himalayas, the Rockies, and the Alps. In this way, continental topography has constantly been renewed, in the face of the erosion that constantly threatens to flatten it.

For the paleontologist, the main implication of plate tectonics is that the geography of the world is constantly changing. Today we recognize seven continents and a host of large islands scattered across the Earth's surface. But 180 million years ago, virtually all of the earth's dry land was assembled into one single supercontinent that geologists call Pangaea ("all lands"). Heat building up below it eventually split Pangaea into two giant continents: Laurasia in the north, and Gondwana in the south. Each of these then fragmented, ultimately to produce the various landmasses that we know today.

During these great movements, climates changed and biological forms shifted. Living populations were isolated or thrown into new states of competition. Species emerged and became extinct, and regions of the world developed their own distinctive assemblages of animals and plants.

ROCKS AND FOSSILS

The rocks that make up the continents of the world come in three different kinds. First there are the igneous rocks, derived from the cooling of molten magma. These include basalts and andesites and tephra ejected by volcanoes on the Earth's surface, and granites that cooled at high depths and pressures, sometimes eventually to be exposed at the surface by weathering. Over the vastness of time, weathering has operated on a grand scale: if you ever find yourself looking at an outcropping of granite, just try imagining that it probably once lay beneath several miles' thickness of rock.

Then there are sedimentary rocks, composed of particles weathered from preexisting rocks before being transported by wind and water, collected, and compacted. Finally, there are metamorphic rocks, which have been reheated enough to flow and recrystallize, as when rough limestone turns to shiny marble.

Fossils are technically any and all traces of past life, not just bones and teeth and shells. Since they are almost exclusively found in sedimentary rocks, these are the only ones we need to dwell on here, except for a passing glance at the volcanic rocks that have proven vital in dating many fossils. When rapidly accumulating sediments cover the remains of dead animals or plants, there is a chance that they will be fossilized. Typically, only the hard tissues such as teeth, bones, or shells undergo fossilization, as their original constituents are replaced by minerals. But occasionally, even soft parts may leave impressions—sometimes amazingly detailed ones—in fine-grained sediments around them.

In the ocean, where sedimentation is relatively continuous, the remains of organisms are routinely trapped in clays, muds, sands, and so forth. On land the process is a bit chancier, and fossils are most often incorporated into the sedimentary record on riverbanks and floodplains, and at the shallow edges of lakes. Such spots also have the advantage—for the paleontologist—of being favorite places for predators to attack prey that have come to drink.

When a terrestrial mammal dies, its remains are likely to be devoured and dismembered by scavengers, and its bones broken and scattered around the landscape. Factors ranging from sun, wind, and water to beetles and bacteria will usually do the rest. If a bone by chance escapes all of these vicissitudes and finds its way to a place of deposition, it will often be further battered en route. This is why most mammal fossils in museum collections are incomplete or damaged in some way, and the most commonly found vertebrate fossils are simply isolated teeth—the hardest tissues of the body.

Occasionally a carcass will be covered by sediments where it lies, and its skeleton preserved intact—naturally enough, the paleontologist's dream. But even this best-case scenario is no guarantee of preservation. As it lies in the rock pile, the fossil must be reasonably undisturbed by earth movements. To be of any use to the paleontologist it has to be uncovered at the surface again by further erosion—where it will be rapidly obliterated by erosion unless it is quickly found and preserved. All in all, a rather chancy proposition, which explains why fossils of many species—especially those species that are thin on the ground in the first place—are rare indeed.

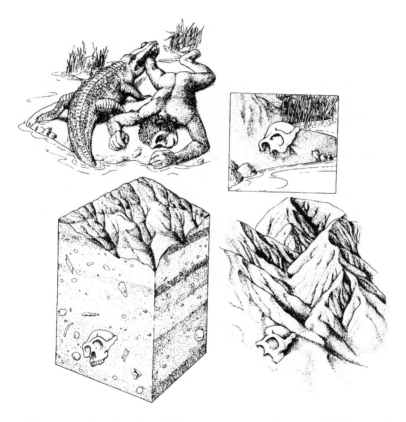

FIGURE 3. The life history of a fossil. After death, most carcasses will be devoured by predators or scavengers (top left). What is left will either weather away or become buried in accumulating sediments (top right). Under the right conditions, mineralization will occur (bottom left). If erosion then wears away the overlying sediments, the fossil will be reexposed at the surface (bottom right), where it must be collected before it is obliterated by the elements.

THE GEOLOGICAL TIME SCALE

For the paleontologist, the most important thing about rocks is the historical record they contain. Ever since the Earth began taking on its familiar form, its continental crust has faithfully registered events happening on local and global scales. Some of this history can still be read, even though much evidence has subsequently been removed by erosion, covered by deposition, or altered by earth movements and metamorphism, sometimes on a gigantic scale.

Once it was established that the Earth was truly ancient, and had not simply been created more or less as we know it today, the first task of the early geologists was to reconstruct the historical sequence encoded in the rocks. This was not easy, for all that the working geologist could see were the rocks that happened to be exposed in any one place. And every local sedimentary basin, let alone each continent, has had its own geological history. Two basic questions thus emerged. One, at the front of every field geologist's mind, was, "What was the sequence of events here?" And the other, usually asked when the geologist had returned home, or had at least struggled as far as the nearest pub, was, "How do I match it up with the sequences we see in other places?"

To approach the first question, early stratigraphers followed two rules. Sedimentary rocks accumulate in piles, one layer atop another, so the first rule was that the sediments at the bottom of any particular pile are older than the strata above. The second axiom was that these layers were originally laid down flat, no matter how earth movements might have tilted or buckled them since. Because most piles of sedimentary rock have undergone at least some deforming and tilting, together with displacement along faults that misalign the layers, stratigraphers first needed to establish the original relationships of the strata. That done, it was time to match up the sequence seen in one place with sequences seen elsewhere.

To some extent, this could be done through lithology—the characteristics of the rock layers themselves. It turned out, though, that this worked only within local sedimentary basins, because each basin has its own geological history. Basins can be large, which is why sheep in southern England graze on the same limestone soils that support the grapevines of Champagne. But every basin has its limits, so stratigraphers found another way to correlate rock formations over broader areas. They recognized certain widely dispersing organisms as "index fossils," characteristic of particular periods. The resulting correlations made possible the development of a standard timescale.

While the succession of major geological periods was basically established by the end of the nineteenth century, means of calibrating that sequence in years are quite recent. As figure 4 shows, Earth history over the last 3.8 billion years is nowadays organized into three major eons that follow the initial period that is informally known as the Hadean, in acknowledgment of the fiery nature of the planet's surface in its earliest days. The first two post-Hadean eons compose the long stretch prior to the earliest fossils known to the early geologists and are grouped together in a larger unit called the Precambrian. The third eon, the Phanerozoic, covers the last 542 million years. Each eon is divided into eras. These are subdivided into periods, which are in turn composed of smaller time units known as epochs.

CHRONOMETRIC DATING

Most current methods of applying real time (in years) to the geological record rely in one way or another on radioactivity. Chemical elements may exist in several alternative forms (isotopes), of which some (the radioactive ones) are unstable: their atomic nuclei spontaneously "decay" to stable states. Conversion takes place at a rate that is constant, measurable, specific to the isotope concerned, and unaffected by environmental factors. Some isotopes decay fast; others more slowly. Chemists express the rate of decay in terms of

Eon	Era	System/Period	Age Myr
Phanerozoic	Cenozoic	Quaternary	0.0118
		Neogene	1.8
			23.0
		Paleogene	65.5
	Mesozoic	Cretaceous	145.5
		Jurassic	199.6
		Triassic	251.0
	Paleozoic	Permian	299.0
		Carboniferous	359.2
		Devonian	416.0
		Silurian	443.7
		Ordovician	488.3
		Cambrian	542.0
Precambrian	Proterozoic	Neo-Proterozoic	Ediacaran 630
			Cryogenian 850
			Tonian 1000
		Meso-Proterozoic	Stenian 1200
			Ectasian 1400
			Calymmian 1600
		Paleo-Proterozoic	Statherian 1800
			Orosirian 2050
			Rhyacian 2300
			Siderian 2500
	Archean	Neoarchean	2800
		Mesoarchean	3200
		Paleoarchean	3600
		Eoarchean	

Cenozoic detail

Era	System/Period	Series/Epoch	Age Myr
Cenozoic	Quaternary	Holocene	0.0118
		Pleistocene	1.8
	Neogene	Pliocene	5.3
		Miocene	23.0
	Paleogene	Oligocene	33.9
		Eocene	55.8
		Paleocene	65.5

FIGURE 4. Simplified International Stratigraphic Chart, showing the major divisions of Earth history with their current datings. On right, Cenozoic chart showing greater detail.

an isotope's half-life—the time it takes for half of the atoms present to decay. Geochronologists have used this property of radioactive isotopes to date rocks containing them.

There are two long-established approaches to such radiometric dating, both first developed in the mid-twentieth century. One embraces accumulation techniques, based on the buildup of stable daughter atoms. The classic accumulation technique is potassium/argon (K/Ar) dating, recently supplanted by the argon/argon (39Ar/40Ar) procedure. Because the half-life involved is very long, these methods and others like them can be used to date very old rocks indeed—volcanic ones are preferred, because when laid down they are heated high enough to purge them of any daughter product, and because they are often found interstratified with fossil-bearing sediments.

The opposite approach is represented by decay techniques, such as radiocarbon (14C), first introduced around 1950. The unstable carbon isotope 14C (radiocarbon) is produced in the upper atmosphere in a reaction governed by cosmic ray influx, and is incorporated into all living things. When an organism dies, it becomes isolated from the carbon cycle, and the 14C it contains begins to decay, diminishing steadily as a proportion of the total carbon present. At 5,730 years, the half-life of radiocarbon is rather short, which means that the method can only be used on samples up to about 40,000 years old. But whereas K/Ar is used to date rocks, 14C has the decided advantage of being able to date fossil specimens directly, provided enough bone protein (collagen) is preserved.

In recent years, the number of approaches to chronometric dating has multiplied, mostly for the fairly recent time frames that are of particular interest to paleoanthropologists. Most of these are "trapped-charge" methods that depend one way or another on the fact that electrons released by radioactivity may become trapped, at a measurable rate, in the lattice structure of various crystals. Good examples are thermoluminescence (TL) and electron spin resonance (ESR) dating.

What Fossils Can Tell Us

Once your dated fossil is sitting on your workbench, you need to extract as much information from it as possible. There are many different ways of going about this, involving specialists of many different kinds. The first step is to determine to what species your fossil belongs—and, if necessary, to create a new species to accommodate it. Then, of course, you need to situate that species in the great Tree of Life. These initial steps are absolutely fundamental to everything else that you do, and they may well prove to be the most difficult steps of all. But only when they are completed should you proceed to what most people regard as the really interesting stuff: reconstructing how your fossil lived back when it was alive, and what role it played in the ongoing soap opera of life.

Apart from its age and the location at which it was found, the most obvious information any fossil has to offer is its morphology—what it looks like. How you are built not only shows to whom you are related, but also determines how you can live. Every species is limited by its structure, both in what it can do right now and in its evolutionary potential for the future.

Of course, when you are confronted with nothing but bones or teeth, it is much easier to reconstruct what their owners might have done in life if you can find a living form whose lifestyle is reflected in features comparable to those of your fossil. The ichthyosaurs, for example, are extinct reptiles whose body form so clearly echoes those of fish that there has never been any doubt that they were swimmers—as is independently confirmed by the marine rocks in which their fossils are found.

On land, the teeth of extinct grazing mammals (and even dinosaurs) are a clear giveaway to their dietary habits, differing as they do from those of their carnivorous contemporaries just as those of carnivores and herbivores differ today. Similarly, the robust forelimbs of digging mammals, or the elongated hind limbs of leaping organisms, were as conspicuous back in the early Cenozoic as they are in the modern world.

The bottom line here is that, even when different animals or plants are not closely related by descent, if they *do* (or did) similar things, they are likely to show similar features as a result of what is called "convergence." A good solution is a good solution, whoever you are. Not every extinct animal has a modern equivalent, so not all past animal behaviors can be inferred from morphologies we see today. Nonetheless, within limits you can reverse engineer to analyze how extinct creatures moved and lived. High-tech methods are also constantly being introduced to help understand the behaviors and diets of extinct creatures. Among such methods is the analysis of certain stable (rather than radioactive) isotopes that are preserved in fossil teeth or bones. By measuring the proportions of different isotopes of carbon, for example, one can tell whether a tropical herbivore was browsing on leaves or grazing on grasses.

Predators preserve an echo of the isotopic ratios of their prey, so they can be included in the calculation, too. Carbon-isotope studies have shown, for example, that some very early human relatives were quite likely eating more meat than had been suspected. Similarly, the further up the food chain you are, the greater the ratio in your bones and teeth will be between the stable nitrogen isotopes ^{15}N and ^{14}N. On this basis, it has been suggested that our close relatives the Neanderthals were highly carnivorous: that, indeed, they may have specialized, at least regionally, in hunting extremely large-bodied prey, such as woolly mammoths and woolly rhinos.

This is merely a sampling of the ingenious approaches that paleontologists have used to flesh out the lives of their long-extinct subjects. But we should also remember that nothing lives in isolation. Every organism belongs to a much larger ecological community, itself a living thing, in which parts cannot be altered without threatening the integrity of the whole. So no matter how much you have been able to infer from an individual fossil that is sitting on the table in front of you, you will never comprehend it completely without understanding the role it played in its wider ecological community.

Helpful here is that fossils rarely occur alone. Instead, whole faunas emerge from fossiliferous rocks. Sometimes they will reflect fairly accurately the larger communities from which they were derived. At other times they are death assemblages, collections of animals that never cohabited in life but were thrown together by the forces of postmortem dismemberment, transportation, and agglomeration. Fortunately, it is usually possible to correct for these postmortem influences to provide a reasonably accurate picture of ancient faunas and habitats. As you go back further in time, and encounter organisms that are increasingly unfamiliar, things become more difficult. But geological and botanical evidence can usually be brought in to help.

To summarize, once you have identified the actors, you can begin to reconstruct the plot of the play—remembering that it was not carefully thought out in advance but was rather a spontaneous drama that unfolded under many different influences. Many of these will have been external, operating entirely without regard to the excellence of the actors' adaptation to their circumstances. Indeed, adaptation turns out to be a two-edged sword. In an erratically changing world, it is often unwise for a population to be too closely adapted to a single environment.

In the Beginning: Life Itself

The question, "What is life?" seems pretty straightforward; the answer is less so. Modern living things are all membrane-bound entities that metabolize (convert energy) and reproduce by means of self-replicating nucleic acids. But the first organisms were probably completely unlike even their simplest descendants today.

In the 1920s it was suggested that complex organic (carbon-based) molecules might have been formed in an oxygen-free atmosphere such as that of primordial Earth, or alternatively in the "hot dilute soup" of the newly formed oceans. But in a breathtakingly prescient letter of 1871, Charles Darwin had already imagined a "warm little pond, with all sorts of ammonia and phosphoric salts, lights, heat, electricity etc. present," in which "a protein compound was chemically formed ready to undergo still more complex changes." In the 1950s, scientists duly generated amino acids, the building blocks of proteins, by sending electric charges through a "prebiotic soup" consisting of molecules of methane, hydrogen, and ammonia in water. Once it was established that the basic components of life could indeed be spontaneously engendered from simple and abundant inorganic precursors, the question became one of the medium in which the transformation occurred.

One leading candidate is the bottom of Earth's late Hadean oceans. In this gloomy setting, warm, alkaline, mineral-rich underwater springs may have reacted with cooler seawater to precipitate out thin films of inorganic molecules consisting of silica, carbonates, clays, iron sulfides, and other minerals. Towerlike edifices of this kind, composed mostly of carbonates, are known today around submarine vents that lie away from the midocean ridges. Such vents furnish much less extreme environments than the scorching hot and highly acidic on-ridge black smokers that support "extremophile" life today. The cooler fluids they emit are well below boiling point, but in the early days the heat gradient would probably have provided enough energy to promote the production of larger, more complex organic molecules, and eventually of self-contained cells.

To be perfectly frank, we don't know exactly how such organic chemicals began behaving like living organisms, or how they developed cellular complexity. Some scientists favor a "genes-first" notion whereby nucleic acids formed at the outset, while others prefer a "metabolism-first" scenario, in which simple metabolic pathways were initially established. However they formed, proto-organisms could have migrated into the sediments of the seabed, to initiate what has been called the "deep biosphere," the mass of microbes that live beneath the sediments of the ocean floor. With the maturing of the Earth's crust, some of these simple organisms would sooner or later have been upthrust into shallow waters, where the penetration of sunlight allowed them to build organic molecules based on carbon dioxide—and photosynthesis began.

This is, of course, just one scenario among several, though whatever the exact details, the basic ingredients for life were present very early on. Sadly, hardly any rocks survive from the Hadean to tell us what happened. The oldest known rocks, about 4.3 billion years old, simply confirm that crust formation began very early. Some 3.8-billion-year-old rocks in Greenland are claimed to contain the waste products of microbial metabolism, but they are pretty thoroughly altered by metamorphism and harbor an active population of cyanobacteria whose activities may have confused the picture. Thus, to begin the paleontological history of life, the history that is documented by physical remains of living things, we have to move beyond the Hadean and into the Precambrian, the period when the sedimentary record reliably begins. The first eon of this almost unimaginably enormous span of time is known as the Archaean ("the Ancient").

THE PRECAMBRIAN (3.8 BILLION–542 MILLION YEARS AGO)

The first unarguable evidence of life comes in the form of Archaean stromatolites, layered mats of sediment up to 3.5 billion years old whose rare modern equivalents are built mainly by photosynthesizing cyanobacteria (blue-green algae). The appearance of stromatolites coincides closely with the earliest evidence of free oxygen in the environment, witness to the fact that they photosynthesized. The advent of free oxygen changed the world, altering the atmosphere from one that was primarily composed of noxious volcanic gases to the one that is vital for almost all life forms today.

The first evidence for free oxygen comes from banded iron formations (BIFs) that began to be deposited about 3.5 billion years ago. The source of most of today's industrial iron, the BIFs were produced as the newly available oxygen began binding to dissolved iron in the seas, producing huge bottom deposits of insoluble iron oxides. When all this chemical activity began, the oceans were a vast "oxygen sink," ready to absorb all of the oxygen that the early photosynthesizers could throw at them—so oxygen-thirsty, indeed, that it took over 1.5 billion years to oxidize all the iron and sulfur they contained. Only when this was achieved, less than 2 billion years ago, did the BIFs cease to form and free atmospheric oxygen begin to rise. Even then a period of red-bed formation followed on land, as terrestrial iron was oxidized.

While all this was going on, the ancient cyanobacteria began to encounter competition from a new form of life: the "true" algae, actually primitive and microscopic photosynthesizing marine plants. The cyanobacteria were probably victims of their own success: as they pumped out oxygen, nitrates (oxides of nitrogen) began to form in the environment—and nitrates were the ideal nutrient for their new eukaryotic competitors, organisms with a cellular structure that included a distinct nucleus and numerous organelles.

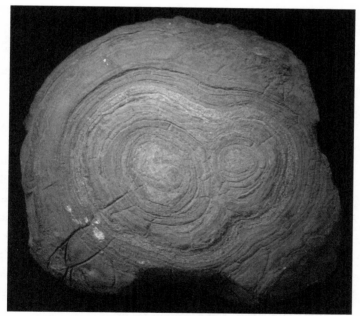

FIGURE 5. *Cryptozoon*, a cabbagelike, reef-forming stromatolite that flourished in Precambrian and Cambrian times. Courtesy of Niles Eldredge.

Soon the stromatolite mats were shared with eukaryotes, and these more complex—though still mainly single-celled—life forms came to dominate the shallow seas of the Earth, building huge thicknesses of limestone as they fixed seawater carbonates. To judge by the increasing variety of stromatolite architectures, this was a time of rapid diversification for the mat builders: they achieved their greatest variety around 1.5 billion years ago and held pretty steady until a gradual decline began about 700 million years ago. The initial proliferation of the new stromatolite makers has been attributed to intense competition for ecological space among the early eukaryotes, and their eventual decline to the rise of grazing eukaryotes that may have exploited them as a source of food. Certainly, today's stromatolites live only in marginal environments where potential predators are unable to thrive.

Still, while stromatolites left the most noticeable evidence of early life during the Archaean, they were not alone. Some early rocks are thought to contain possible microfossils as well as biomarkers, chemical by-products of metabolism. Whether the metabolizers concerned were equivalent to any bacteria surviving today is debated, but they do seem to provide independent evidence of life at the very beginning of the Archaean.

Eukaryote fossils do not make their appearance until long after the beginning of the Proterozoic ("before life") Eon some 2.5 billion years ago. The first of them are rather unimpressive, basically amounting to tiny acritarchs ("of uncertain origin") that probably represent microscopic floating algae. These tiny things have been a durable feature of the fossil record ever since their appearance some 1.4 billion years ago. But not until the end of the Proterozoic, in the Ediacaran Period, do we find fossils we can truly recognize as those of animals and evidence of something that we can really call a "fauna." Meanwhile, though, the Earth itself had been through some quite traumatic times.

Snowball Earth
Geologists call the period preceding the Ediacaran the Cryogenian ("frozen"), and for as good a reason as the name of the first

stage of Earth's history evokes Hades. At about a billion years ago, the Earth's continental masses came together in a single supercontinent that has been called Rodinia. With very extensive exposure of dry land, increased weathering robbed the atmosphere of carbon dioxide, which abundant tropical rainfall converted to carbonates and ultimately returned to the sea. This decline in the principal greenhouse gas set the stage for climatic cooling. Ice caps formed at the poles and advanced toward the equator. Eventually, the Earth's bright frozen mass reflected so much of the Sun's heat that the cooling became a runaway process. The result was a "snowball Earth" on several occasions between about 720 million and 630 million years ago.

The pure version of this scenario encases the Earth entirely in ice during these periods of intense cold, bringing life on the planet almost to an end. The fact that life survived has led some to prefer a "slushball Earth" idea, whereby the ice cover was never quite complete. Either way, widespread geological evidence for low-latitude glaciations supports some form of the snowball Earth hypothesis, and there is general agreement on a series of Cryogenian global cooling crises that set the stage for the future evolution of life on Earth.

What is also apparent is that, when it came, the end of the cycle was sudden. Tectonic activity didn't stop just because of the ice. Beneath its frosty veneer Rodinia was busily breaking up, and there is no doubt that volcanic vents and fissures were energetically pumping fresh carbon dioxide through the ice and into the air. With no rain to wash them out of the atmosphere, the greenhouse gases accumulated and at the end of several million years of snowball Earth, the concentration of atmospheric carbon dioxide would have risen hugely. The deglaciation would have been as self-reinforcing as the initial cooling had been: water falling on the newly exposed landscape as acid rain would have intensified weathering and the transport of nutrients to the formerly stagnant seas, causing cyanobacteria to bloom. Their photosynthesis led to a rise in atmospheric oxygen, as witnessed by the brief return of banded iron formation as volcanically derived iron was oxidized.

The exact extent to which life was banished by the Cryogenian glaciations is uncertain, but clearly it suffered a near-fatal trauma. As a result there is little fossil evidence beyond the acritarchs to document this period. But while before the second major Cryogenian glacial episode at around 630 million years ago, recognizable acritarch species had lingered for as much as 100 million years, afterward their longevity was drastically reduced, suggesting that something new was stirring ecologically. One possibility is that, in a warming world, acritarch algae had for the first time encountered predators.

The presence of predators meant that organisms were no longer limited solely by the amount of nutrition available. Ecological conditions had become more complex, and evolution had found a new stimulus. The very earliest animal fossils, putatively bilaterian embryos from China, are possibly up to 599 million years old. They suggest that, in the immediate aftermath of snowball Earth, the world was radically changing.

The Ediacaran
(630 Million–542 Million Years Ago)

The record of complex life begins essentially with the impressions of soft-bodied animals belonging to a fauna known as the Ediacaran. In many ways it's amazing that these creatures should have been preserved at all. Nowadays the shallow seafloor is thoroughly reworked by the numerous organisms living down there. But in the Proterozoic, the mortal remains of the fragile Ediacaran beasts could rest in peace on the microbial mats on which they had lived. Dating between 575 million years ago and the end of the Precambrian some 33 million years later, the Ediacaran biota has excited enormous debate. Its members have been interpreted together as an entirely separate evolutionary development, and individually as the precursors of several lineages that flourished in later

times. Whatever the case, the Ediacaran forms were clearly mul-
ticelled animals with distinct body plans. Some of them at least
were mobile. They were also apparently diverse. They ranged from
wormlike forms identified from casts of their burrows, through dis-
coids sometimes identified as jellyfish; from apparently segmented
organisms and somewhat tunicate-like, mud-filled bags of uncertain status, to quilted or frondlike creatures.

At one time these last were thought related to today's sea pens,
opening the way to construing many Ediacaran forms as precursors of living groups. However, the current trend is in the opposite
direction, and one authority has classified the classic Ediacaran
forms together in their own phylum Vendobionta, a self-contained
group of altogether uncertain affinities. If this is the case, the core
Ediacaran biota was a natural experiment that ultimately failed,
leaving no descendants in the post-Precambrian world. Still, this
is controversial, and molecular clock estimates suggest that ancestors of today's metazoans were already in existence 600 million
years ago.

Even the term "Ediacaran biota" is ambiguous. Sometimes
applied just to the bizarre vendobionts, at other times it embraces
all organisms of the Ediacaran Period. In the latter sense the Ediacaran biota anticipates later times: plenty of evidence exists that
bacteria of various kinds continued to flourish in the oceans, as
did protists, red and green algae, sponges, shelled organisms, and
worms. However, they flourished in a world that did not closely
resemble the one to come. Many Ediacaran forms probably grazed
on the algal mats produced by the stromatolite microbes and, as
predators, may even have played a role in the decline of stromatolite abundance that began at about the time they made their
appearance.

Some small shelly fossils from the Ediacaran, including the tube-
like Cloudina, bear holes apparently bored by predators, implying
some increasing ecological complexity even if we don't know who
the predators were. Still, few Ediacaran organisms made it into the

FIGURE 6. *Dickinsonia,* a probable animal that is one of the classic Ediacaran fossil forms. Courtesy of Niles Eldredge.

Phanerozoic Eon, the age of "revealed life." In the first major extinction on record, they were replaced by the earliest members of the "Cambrian Explosion," who played the ecological and evolutionary games in entirely new ways.

R. J. Berry
Our Ecological Systems

 IN HIS BOOK *Ecology and the Environment*, Berry shows how the Earth behaves like a "green machine," driven by the Sun (directly and indirectly) and providing us with "ecological services." Through history, humans have viewed nature with awe, but also exploited it rapaciously for survival, bringing us to a time when we must understand these ecological services better, and use them more carefully. In that light, Berry surveys the rise of ecological science, showing its increasing importance. The sheer weight of human numbers and use of resources is forcing us to watch our "ecological budget" in a way unnecessary in past ages. Science provides data for compiling this budget, which is made up of complex interactions between the Earth's structure, geological processes, climate, chemistry, and biology, which includes species and their niches. Some scientists have gone even further, suggesting that the Earth might even be a massive self-regulating organism. However we look at it, Berry argues that a sensible budget involves human "stewardship." As the proverbial "naked

ape," humans can behave as the "most dangerous species" on Earth, or acknowledge a spiritual role, which includes working with nature. Part of having ecological literacy, Barry argues, will be in seeing how transcendent (or religious) values can guide human treatment of nature. He calls it a seamless alliance between "wise science and wise faith."

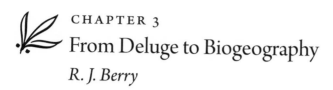

CHAPTER 3

From Deluge to Biogeography

R. J. Berry

THROUGHOUT MOST OF human history, we had no reason to doubt that we were living in a world unchanged from its beginning, and according to Plato writing four centuries before Christ, a world that is not only unchanged, but virtually unchangeable. From this belief to an assumption that everything exists for the benefit of human beings is only a short step, an expectation natural enough in itself and seemingly in accord with the Bible. This meant that for many centuries there seemed to be no need to study the natural world for its own sake; knowledge about it came either from the information necessary to survive or from philosophical speculation about the ways of the Creator.

This produced a brake on our perception, well illustrated by the criticism of Harvard biologist Ernst Mayr:

> Plato's dogmas had a particularly deleterious impact on biology through 2,000 years. One was essentialism, the belief in a constant *eide* separate from and independent of the phenomena of appearance. The second was the concept of an animate cosmos, a living harmonious whole, which made it so difficult in later periods to explain how evolution could have taken place, because any change would disturb the harmony. Third, he [Plato] replaced spontaneous generation by a creative power. A demiurge. And fourthly [was] his great stress on "soul."

This Platonic constraint began to lessen in the seventeenth century; indeed, by its end, discoveries from astronomical observations and geological processes led to belief in changelessness becoming increasingly untenable. The existence of deep time and deep space became widely accepted. This shift in perception was necessary for science—never mind ecology—to develop. It was not so much a denial of our heritage as a reinterpretation of the information on which it was based in the light of new discoveries. We began to recognize that the world is not here solely for our benefit and that it is bigger, grander, and older than our forebears ever suspected.

A key actor in this story was an English naturalist, John Ray (1627–1705), sometimes called "the father of natural history." Ray argued that nature can be studied for its own sake and on its own terms, and that it exists for reasons of its own, not just to serve humans. He was certainly not a natural iconoclast. He was the son of a blacksmith, born in a small village in eastern England, but he seized the opportunities available to him and challenged the traditions he inherited, particularly in exercising his faculty for observation and description. He lived in a time of social and intellectual turbulence—the flowering of the Age of Reason, a period that included Galileo's debates in Rome as well as the philosophical challenges of John Locke and David Hume. During Ray's lifetime, agricultural improvements were gathering pace; when he was middle-aged, the rapidly growing London was almost destroyed in a Great Fire (1666) and then rebuilt under the genius of Christopher Wren. He was an early fellow of the Royal Society of London, being admitted in 1667, seven years after the Society's founding; he shared in the adolescence of the reformed Church of England, resigning his fellowship of Trinity College, Cambridge (where Isaac Newton was one of his contemporaries) because of his reservations about the 1660 Act of Uniformity, which required all citizens to adhere to an official creed.

During the last quarter of the seventeenth century, Ray published under his own name, or that of his pupil and patron Francis Willughby, classifications of the major groups of animals and plants, taking a major step toward a natural system and preparing the way for Linnaeus, both in terms of method and of data. The French biologist Baron Cuvier described Ray as "the first true systematist of the animal kingdom . . . His works are the basis of all modern zoology." Gilbert White of Selborne regarded Ray as his mentor, scientifically and theologically. He extolled Ray as "the only describer that conveys some precise idea in every term or word, maintaining his superiority over his followers and imitators, in spite of the advantage of fresh discoveries." More soberly but perhaps more importantly in recognizing his legacy was the clarification in his book *Ornithology* (1678) that "what properly relates to Natural History," involves "wholly omitting Hieroglyphics, Emblems, Morals, Fables, Presages or ought else appertaining to Divinity, Ethics, Grammar, or any sort of Human Learning." By separating the study of nature from speculative religious concerns, Ray liberated the study of the natural world from the specifically religious constraints of biblical interpretation and opened the way for ecology; Ray followed the Protestant Reformers in seeing the book of scripture as free of allegory, to be interpreted only in a literal and historical sense. Put another way, Ray applied the implications of the belief—well-established in his time—that God had written a book of nature as well as a book of scripture. This belief in two separate books showed the irrelevance of using religious allegory to explain nature. The book of nature stood on its own. He declared in his book *The Wisdom of God Manifested in the Works of Creation* (1691):

> It is a generally received opinion that all this visible world was created for Man [and] that Man is the end of creation, as if there were no other end of any creature but some way or other to be serviceable to Man. . . . But though this be vulgarly received, yet wise men nowadays think otherwise.

Ray's statement provided a motive for the study of the natural world for its own sake. It also challenged the dominant assumption of his age that, under God, all things are subordinate to human beings.

The decades after Ray's death saw the use of reason to explore the nature of God reach a climax. The beginning of the eighteenth century was the heyday of "physico-theology," epitomized by the Boyle Lectures, founded with a bequest from the chemist Robert Boyle (1627–1691) to be delivered in London churches and directed against unbelievers. A particularly notable series, given in 1711–1712 by William Derham, was titled *Physico-theology or a Demonstration of the Being and Attributes of God from His Works of Creation.* Derham based his arguments on Ray's *Wisdom,* acknowledging "my Friend, the late great Mr. Ray."

Physico-theology treated God as the First Cause, the Divine Watchmaker who had created all things and then retired above the bright blue sky. It reached its climax in the writings of Archdeacon William Paley (1743–1805), particularly his *Natural Theology* (1802), which plagiarized Ray's examples in the *Wisdom* without acknowledgment. Although the two books have much in common, they are poles apart: Ray worshiped God for his marvelous works in nature (theism); Paley saw nature as the proof of God's handiwork (deism). The last major manifestation of this traditional version of natural theology was the Bridgewater Treatises, endowed by the Reverend Francis Egerton, eighth Earl of Bridgewater, who (allegedly in expiation for a misspent life) charged his executors, the archbishop of Canterbury, the bishop of London, and the president of the Royal Society of London to pay eight scientists one thousand pounds each to examine "the Power, Wisdom, and Goodness of God, as manifested in the creation; illustrating such work by all reasonable arguments." Their books were published between 1833 and 1837.

Interest in natural theology—extrapolating theology from science—had social benefits, turning the attention of ordinary people to a study of the natural world. The mid-nineteenth century saw crazes for drawing-room aquaria and for growing exotic ferns. A survey in 1873 revealed 169 local scientific societies in Britain, of which 104 claimed to be field clubs. Most of these had come into being since 1850, a rate of formation of 10 per year. More and more people were studying the natural world—not as hunters or farmers but as inquirers.

But knowledge of the world was moving on; traditional natural theology was dying. The early geologists were providing more and more evidence of a long Earth history. Alongside their understanding of "deep time," it was becoming apparent that the universe was vastly bigger than envisaged by medieval astronomers. The telescopes of William Herschel (1738–1822) and others showed that the solar system itself was moving through space. There is "deep space" as well as "deep time." At the same time that Chalmers, Kidd, Whewell, Bell, Roget, Buckland, Kirby, and Prout were laboring over their Bridgewater Treatises, Charles Darwin was sailing around the world on the HMS *Beagle* (1831–1836). The outline of his evolutionary ideas were in place by early summer 1837. Then in 1844 Robert Chambers' *Vestiges of the Natural History of Creation* was published. It was effectively a tract against Paley's version of deism. The *Vestiges* was an immediate bestseller. In its first ten years it sold more copies than Darwin's *Origin of Species* did fifteen years later, yet it was full of errors. For Darwin, "The prose was perfect, but the geology strikes me as bad and his zoology far worse." Nevertheless it stirred debate. Darwin welcomed it on the grounds that "it has done excellent service in calling in this country attention to the subject and in removing prejudices."

Chambers wrote that when there is a choice between God's direct "special creation" of things in nature and the operation of general laws instituted by the Creator, "I would say that the latter is greatly preferable as it implies a far grander view of the divine power and dignity than the other." Since nothing in the inorganic world "may not be accounted for by the agency of the ordinary forces of nature," why not consider "the possibility of plants and animals having likewise been produced in a natural way"? In other words, he sought to supplement natural theology by verifiable natural law(s). William Whewell, one of the Bridgewater authors, had argued similarly, but in less accessible prose. Darwin quoted Whewell on the page opposite the title page of *The Origin of Species*: "But with regard to the material world, we can at least go so far as this—we can perceive that events are brought about by insulated interpositions of Divine power, exerted in each particular case, but by the establishment of general laws."

Chambers was reviled from many quarters. Adam Sedgwick, professor of geology at Cambridge and geological mentor to Charles Darwin, lambasted the *Vestiges* in an eighty-five-page diatribe in the *Edinburgh Review*. He wrote to his friend Charles Lyell, "If the book be true, the labours of sober induction are vain; religion is a lie; human law is a mass of folly and a base injustice; morality is moonshine; our labours for the black people of Africa were works of madmen; and man and woman are only better beasts."

Sedgwick firmly identified himself with the old view of the natural world—an unchanging order ruled by a Creator who mandated a stable organization and social manners. He wrote: "The world cannot bear to be turned upside down; and we are ready to wage an internecine war with any violation of our modest and social manners. . . . It is our maxim that things must keep their proper places if they are to work together for good."

THE FLOOD

Like it or not, the world *was* in the process of being turned upside down. Within a few years Darwin would comprehensively scuttle the restricted deist notion of God as being nothing more than a clever designer. The Bible was having to be reinterpreted. Copernicus and Galileo had moved the Earth from the center of all things to that of a planet circulating around the Sun. Where was heaven in this new cosmology? And what about the age-old stories of God coming in judgment through natural disasters, like the flood?

Thomas Burnet (1635–1715) wrote *The Sacred Theory of the*

Earth between 1684 and 1690. It appeared alongside two other great works published at the same time: Ray's *Wisdom* and Isaac Newton's *Principia Mathematica* (1687). Although Burnet's *Sacred Theory* was the most idiosyncratic of the three (it has been said that Burnet would have become archbishop of Canterbury if he had not written it), and has been the least enduring, it was the most influential in its time.

The *Sacred Theory* was an attempt to explain the facts of geology—why the surface of the Earth is so uneven in terms of mountains, seas, and landmasses, and why islands exist. Burnet began with the almost unquestioned assumption of his time that God had created the world as a series of concentric layers, with a crust lying over water, and "no Rocks or Mountains, no hollow caves, nor gaping Channels, but even and uniform all Over." Rivers ran from the poles to the tropics, where they dissipated. This primitive order disappeared in the devastation of the biblical flood. For Burnet, the Earth of his time was the shattered ruin of a "perfect" preflood creation; the oceans were gaping holes and the mountains upturned fragments of the old Edenic crust. He was particularly offended by mountains: "If you look upon a heap of them together they are the greatest examples of confusion that we know in Nature; no tempest or earthquake puts things into more disorder." Burnet's book greatly impressed Newton, who wrote to Burnet to congratulate him. The book was reprinted repeatedly throughout the 1700s and was regarded as a significant geological text.

The flood is, of course, tied in the Bible to the story of Noah, who took into an ark built at God's command "a male and a female of all beasts, clean and unclean, of birds, and of everything that creeps on the ground" to save them from drowning in a great deluge. The interpretation of this story has exercised biblical exegetes over many centuries: Could every species fit into the ark? Where was the fodder stored? What happened to the dung? How could carnivores and herbivores coexist? A new level of questioning arose in the seventeenth century, however, with increasing knowledge of the New World. The Americas were variously interpreted as a land that had wholly escaped from the flood, a sodden continent only recently risen above the waters, or the New Atlantis that Plato described in his dialogue *Timaeus*.

Even British jurists tried their hand at explanations. Lord Chief Justice Matthew Hale (1609–1676) proposed that American animal life had appeared through a kind of migration and subsequent degeneration—or, if not, perhaps that American species are worldwide and they would be found elsewhere when Africa and Asia were more thoroughly explored. He interpreted native Americans as descendants of the lost tribes of Israel, naked innocents who had escaped the fate of Noah's wicked generation, or perhaps degenerate savages who had wandered away after the ark grounded. In France, Isaac de La Peyrère (1594–1676) argued that the flood was a local event and that Adam was not the first-created man, but merely the first Jew, with the natives of the Americas representing other species. Hale rejected this as "immoral and irreligious" and insisted that all humans were descended from Adam, and that the same argument could apply to the animal kingdom: all animals were descended from those that came out of the Ark.

A century after Hale and La Peyrère, Linnaeus listed fourteen thousand species, around a third of them animals. How could so many creatures have been passengers on the ark? Linnaeus regarded himself as "the publisher and interpreter of the wisdom of God" and opted for a revisionist version of the Bible. He suggested that all living beings, including humankind, originated on a high mountain at the time the primeval waters were beginning to recede. Extrapolating backward, as it were, he concluded that in the beginning only one small island had been raised above the surface of a worldwide sea, which must have been the site of Paradise and the first human home.

As the mountain grew higher and higher, Linnaeus proposed that it could have presented an increasing range of ecological conditions, arranged as belts from tropical to polar zones. He envisaged organisms moving in turn to the latitudes where they were to

FIGURE 7. Noah's ark from the *Nuremburg Chronicle* (1543). Reconstructions of the ark from biblical data became increasingly strained in the sixteenth century as travelers began to bring to Europe a wealth of unusual creatures from newly discovered regions.

remain for the rest of time. He did not contest the biblical account of creation, but questioned the ark episode; it played no part in his understanding. He asked, "Is it credible that the Deity should have replenished the whole earth with animals to destroy them all in a little time by a flood, except a pair of each species preserved in the Ark?" He believed it was a combination of sheer chance, the speed of migration, and the helping hand of God that determined exactly where each species ended up.

Under Linnaeus' scheme, the whole system was delicately balanced, with plants and animals perfectly constructed for the environment where they lived. Linnaeus saw the natural world as a static structure, with every tiny detail preordained by the Creator God. Unfortunately for his argument, this thinking opened the door for more radical accounts. If every species was so well suited to its environment, it was difficult to imagine their members traveling across vast tracts of land from Mount Ararat. Perfect adaptation of organisms seemed incompatible with migration from a single source.

A German zoologist, Eberhardt Zimmermann (1743–1815), was

ruthlessly destructive of the Linnaean interpretation. He pointed out that the first pair of lions on Linnaeus' mountain would soon eat the first pair of sheep, then the goats, cows, llamas, buffaloes, zebras, and so on, in quick succession. Finally the lions would die from hunger. Zimmermann wrote that it would be far better that every animal should be created in the area where it was destined to live, under the same climate that it now enjoys, and with the same food rations already in abundant supply.

Once the improbability of every animal coming together in the ark or having a common source on Mount Ararat was recognized, it became plausible to probe into the origin and distribution of organisms. The French Comte de Buffon (1707–1788) pointed out that there were regional associations of animals, inhabiting different areas of the globe; the Swiss botanist Augustin de Candolle (1778–1841) defined botanical "regions" where endemic (that is, native) species were found. Faunas and floras of different areas started to be written.

The idea of multiple centers of creation began to be canvassed. Increasing knowledge of the distribution of animals and plants was a major factor in leading to the recognition that living things required investigation in their own right—a major step to study of the real world, as opposed to castles in the mind like Burnet's scheme. Wonderment at nature and a passion to collect animals and plants sparked the revolution that led to what we now call ecology.

BIOGEOGRAPHY: DISTRIBUTION OF ANIMALS AND PLANTS

In the mid-eighteenth century, Carl Linnaeus set a trend for biological realism with his collections, derived from systematic soliciting of specimens from a wide range of localities. He sent out so-called apostles to many parts of the globe. The most famous was Daniel Solander (1733–1782), whom Linnaeus intended to be his successor. Solander, however, having gone to England to publicize Linnaeus' ideas, stayed there. Eight years after his arrival (in 1768), he was employed to accompany Joseph Banks (1743–1820) on the first voyage of the *Endeavour* under Captain James Cook. Banks bestrode British science for half a century. He was president of the Royal Society for over forty years, unofficial director and reinvigorator of the Royal Botanic Gardens at Kew, patron of many scientists and scientific expeditions, and energetic advocate of transplanting economically important species.

Specimens that Banks collected on the *Endeavour* voyage led to the recognition of 110 new genera and around 1,300 new species, but his main legacy is as an administrator rather than a practicing biologist. A younger German contemporary, Alexander von Humboldt (1769–1859) had a much greater influence on science sensu stricto. Humboldt traveled extensively in Central and northern South America, fired by the enthusiasm of his University of Göttingen friend, Georg Forster (1754–1794), who had been a naturalist on James Cook's second voyage (taking the place of Banks, whose demand for accommodation was so excessive that he was refused a place on the ship).

During his travels, Humboldt made observations that form the basis for quantitative ecology, physical geography, and meteorology. On his way across the Atlantic, he spent some time on the Canary Island of Tenerife where he described vertical zonation in the vegetation of the central mountains, the sort of analysis that today forms the preliminary stage of any ecological study. He invented isobars and isotherms as an aid to showing the limits of particular species and natural assemblages, and developed botanical arithmetic, the ratio of species in one group of plants to that in another. This ratio can show the predominant forms present in a region and the general relationships between different groups. The technique was developed further by Augustin de Candolle (1778–1841), and used by Humboldt's friend Christian von Buch (1774–1853) in his essay on the Canary Islands, which for the first time set out evidence for geographical speciation; by Charles

Darwin in comparing the flora of different archipelagos; and by Joseph Hooker (1817–1911) in comparing continental and insular populations.

All of this led to a new science: biogeography. Its founding is generally attributed to the French scholar the Comte de Buffon a few decades earlier. Buffon was mainly a laboratory scientist, but his study of regional associations of animals showed the way forward. That point notwithstanding, the major stimulus for biogeography came from Humboldt and the pioneering scientific voyages of the eighteenth and nineteenth centuries.

The early voyages were concerned almost exclusively with exploration and mapping. From the biological point of view, the first significant one was James Cook's first voyage in 1768–1771, during Buffon's lifetime; Cook took with him Joseph Banks and Linnaeus' disciple, Daniel Solander. The expedition tends to be remembered for Cook's discoveries in Australia and for Banks' collections, but its overt purpose was to establish on Tahiti a station to observe a transit of Venus across the Sun, allowing the distance of the Earth from the Sun to be calculated and hence the calibration of astronomical data critical for navigation. It was followed by a French expedition in 1800–1804 under the command of Nicolas Baudin (1754–1803), which provided a mass of material for Lamarck's studies in Paris.

More important, because of the generalizations that resulted from it, was the voyage of HMS *Beagle* under Robert Fitzroy in 1831–1836, with Charles Darwin as Fitzroy's "gentleman-companion."[1] Like Ray before him, Darwin was entranced by the scenes he encountered: "The sense of sublimity which the great deserts of Patagonia and the forest-clad mountains of Tierra del Fuego excited in me, has left an indelible impression on my mind."

Darwin was primarily a naturalist; he was not a theoretician nor was he a natural contrarian. He marveled at the lushness of the tropics that he first encountered in Brazil. He was impressed by the biota of South America and how it was apparently adapted to local conditions while differing from that on other continents, and how one form replaced another along the length of South America as conditions changed. He was puzzled about the numbers of unique species on oceanic islands. It was only after his return to Britain that he realized the significance of the variety of birds on the Galapagos Islands, but he noted at the time, "I never dreamed that islands about fifty or sixty miles apart and most of them in sight of each other, formed of precisely the same rocks, placed under a quite similar climate, rising to nearly equal height, would be differently tenanted. . . ."

Darwin's voyage (he never left Britain again) was followed a few years later by that of Joseph Hooker on HMS *Erebus* (1839–1843), sent to search for the site of the South Magnetic Pole. The young Hooker idolized Darwin; he took a copy of Darwin's *Voyage of the Beagle* with him on the *Erebus*. Later in life he wrote:

> The [book] impressed me greatly, I may say despairingly, with the genius of the writer, the variety of his acquirements, the keenness of his powers of observation, and the lucidity of his descriptions. To follow in his footsteps, at however great a distance, seemed to be a hopeless aspiration; nevertheless they quickened my enthusiasm in the desire to travel and observe.

ISLANDS AND THEIR SPECIES

Good scientist that he was, Hooker was influenced by his observations. His notes on the voyage quickly expanded from the minutiae of collecting to questions of geographical distribution. Madeira, his first island, "strongly reminded me of some of the islands on the West of Argyllshire [off Scotland] . . . The ravines are quite like Scotch ones, but more sparingly wooded." In a letter to his father written during the journey through the South Atlantic, he clearly thought that the island biotas he would find would be determined by temperature.

By the time the *Erebus* reached Kerguelen Island in the southern Indian Ocean, midway between Africa, Australia, and the Antarctic, Hooker had begun to ask deeper questions about the relation between the flora of islands and continents. He concluded that the most marked influence on the Kerguelen flora was "Fuegian," reminiscent of the flora of Tierra del Fuego at the tip of South America. As he journeyed on, he came to see the Fuegian flora as "the great botanical centre of the Antarctic Ocean"; all the islands south of New Zealand, together with the Falkland Islands, South Georgia, Tristan da Cunha, and Kerguelen "seemed to have borrowed plants" from there. He found this astonishing; Kerguelen, for example, was five thousand miles from Tierra del Fuego. Not only that, Fuegia possessed a great number of English plants. His mind was drawn to "that interesting subject—the diffusion of species over the surface of the world."

This led to a long-continued debate with Darwin: Were oceanic islands the vegetated relics of now-submerged continents, or did their flora arrive by migration, as Darwin believed? This question, of course, is the heart of biogeography; it is the same problem Zimmermann faced in objecting to Linnaeus' idea of a spread from Ararat. We now know that both mechanisms are operating. Some of the best data on this point come from the colonization of recently erupted volcanoes.

Colonization is easier and turnover more rapid when empty habitats are available—this happens when a volcano forms a new island or sweeps an existing area of all its life. Some wide-ranging species quickly establish themselves whenever an opening occurs; they have been described as supertramps—good colonizers but usually poor competitors, a distinction that recalls that between *r*- and *K*-strategists.[2] The island of Krakatau in Indonesia exploded in 1883. In the first phase of reestablishing of biological life, most of the colonizers away from the shore (an average of 2 species a year) were windborne, although nearly as many (1.64 species per year) colonized the shoreline. In the early years, only an average of 1 animal species every seven years penetrated into the interior. During the next phase (1897–1919), windborne pioneers lost ground, and animal penetration of the interior increased tenfold (to an average of 1.32 species per year) as fruit-bearing plants grew up and provided food for immigrants.

Then in 1930 further volcanic activity produced a new island, Anak Krakatau ("Child of Krakatau"), which has experienced sporadic eruptions ever since. A visitor six months after its appearance wrote:

> Its virgin shore [is] composed entirely of dark grey ash, black cinders and white pumice stone. No plants would grow here until weathering and bacteria had had time to create soil in a year or two, but seeds, along with debris like banana stems and other vegetable matter, were awaiting their time to establish themselves. . . . The only abundant insects were scavengers which could feed on whatever plant life the sea brought them—a springtail, a beetle, a species of ants, a tiny leaf-mining moth and a mosquito.

The first birds seen were a Pacific reef-egret and a beach thick-knee, together with migrants like the common sandpiper, grey wagtail, Pacific golden plover, Mongolian plover, whimbrel, and great knot. In 1985, seventy-two species of small flying creatures were collected in ten days in plastic containers filled with seawater in places where the lava had flowed—a constant rain of arthropods. A great increase in bird species occurred in the same year, coinciding with the first fruiting of the island's figs and the consequent opening of a new "habitat window." As these immigrants established themselves, some of the earlier-arrived ground-nesting species disappeared. Then raptors came to feed on the fruit-eating species. About 150 plant species now inhabit the island. The same process has been occurring on Surtsey, an island produced by a volcanic eruption off the southwest corner of Iceland in 1963, although here the land surface has been eroding so quickly that the succession has not been as marked as for Krakatau (see figure 8).

One can also study colonization experimentally. Ed Wilson and Dan Simberloff painstakingly undertook such a study, fumigating four small mangrove islands off the coast of Florida so as to kill all the resident animals and then monitoring the islands' recolonization over a period of years. Intriguingly, a similar pattern of colonization to that following volcanic eruptions occurs in situations that are intuitively stable, such as mature forest or rocky seashore. University of Washington ecologist Bob Paine has followed the sequence of events after an old tree has fallen or a patch of shore has been artificially cleared of its animals and plants. He found that the dynamics of the reestablishment of the fauna and flora are the same as in the more spectacular circumstances of Krakatau and Surtsey, albeit in a less obvious manner.

A major advance in the understanding of island biotas has been "the theory of island biogeography" that Robert MacArthur and Ed Wilson put forward in 1963 and expanded into a book in 1967. The core of their thesis was that a balance exists between immigration to an island determined by its distance from the mainland and extinction of local populations, which will vary with the island area. In other words, the number of species on an island will be the difference between those continually reaching it and those that are being lost (see figure 9). Their insight was that the species composition is a present dynamic, not a simple historical event. Species are continually going extinct locally; species are continually appearing and establishing themselves.

What MacArthur and Wilson expressed as a quantitative theory is really something naturalists have long intuitively known. English naturalist Alfred Russel Wallace, codiscoverer with Darwin of the idea of natural selection, described the same outcome in his *Island Life* (1881):

FIGURE 8. (a) Surtsey: An island in the Westmann group off the south coast of Iceland formed in 1963 as the result of an undersea volcanic explosion (Photo: Sturla Fridriksson). (b) Pioneer shore community of lyme grass and sea sandwort on sandy lava on Surtsey. These species may have come from a neighboring island or the mainland of Iceland (Photo: Borgthor Magnusson).

> The distribution of the various species and groups of living things over the earth's surface and their aggregation in definite assemblages in certain areas is the direct result and outcome of . . . firstly the constant tendency of all organisms to increase in numbers and to occupy a wider area, and their various powers of dispersion and migration through which, when unchecked, they are enabled to spread widely over the globe; and secondly, those laws of evolution and extinction which determine the manner in which groups of organisms arise and grow, reach their maximum, and then dwindle away. . . .

British botanist H. C. Watson put numbers on the process. He pointed out that a single square mile of the English county of Surrey held half the plant species found in the 650 square miles (1700 km²) of the county as a whole. Henry Gleason generalized this in a 1922 paper "On the Relation between Species and Area." Then, in the 1950s, Philip Darlington applied the same principle to islands. He noted that in a range of islands, when the area was increased by tenfold, the number of species merely doubled. He also found that the species present in any one place are not fixed; some animals or plants disappear while others appear. All of these studies point to the same scenario: a coming and going of species, and a tendency for them to aggregate in certain places.

It was left to MacArthur and Wilson to give formal expression to these ideas in the 1960s. They suggested that recurrent colonizations and extinctions create the appearance of an equilibrium, in which the number of species remains relatively constant although the species concerned vary over time (see figure 9). MacArthur and Wilson used data from the recolonization of Krakatau to support their thesis.

Data about extinction and recolonization are more accurate for birds than for other groups, because it is usually easy to record new breeders or species failing to breed. The reason for failure may be

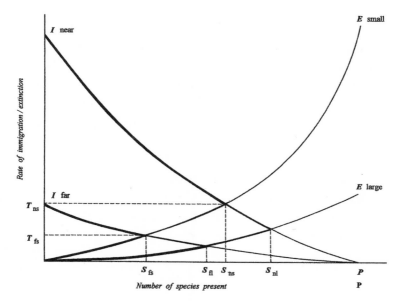

FIGURE 9. The number of species on an island (S) is the result of a turn-over (T) between rates of immigration (I) and extinction (E). Immigration is affected by the proximity of an island to its source of colonizers (n = near, f = far) and the size of the island (s = small, l = large) (based on Robert H. MacArthur and E. O. Wilson, *Theory of Island Biogeography* [Princeton, NJ: Princeton University Press, 1967]).

because numbers are declining generally or because of local factors like competition, change in habitat, or even pure chance. One complication is that different species have different mobility, whatever their potential powers of flight. For example, woodpeckers rarely cross water, although obviously they may be involuntary colonizers through accidents of weather or other rare events. The ancestors of the finches on the Galapagos or the honey-creepers on Hawaii cannot be said to have "intended" to settle where they ended up, and their successful establishment and breeding represent extremely unlikely events. It is impossible to know how many of their relatives perished at sea, although some indication is given by twitchers' delights—"vagrant" birds from distant parts of the world that appear on distant shores to the excitement of fanatical bird-watchers ("twitchers")—albeit mostly to die a lonely death.

MacArthur and Wilson described the balance between immigration and extinction as a state of equilibrium. In fact, it is really nothing more than a logical necessity. The number of species on an island can only be increased by two processes—immigration, which in turn depends on the distance of the island from the source of potential colonizers and the availability of ecological space for them—and be decreased only by those that fail to survive, that is, by extinction. Certainly, the theory needs supplementing with ecological information. For birds, the idea of optimal foraging improves the theory's predictions, since organisms are more likely to stay longer in an area if the distance to a neighboring island is great.

GENETIC VARIATION: THE FOUNDER EFFECT

In the 1970s and 1980s, a major change in population biology took place. One of its effects has radically affected the way we think of the way species that are native to an area—called *endemic species*—may originate. Biologists traditionally thought of individual animals or plants as genetically rather uniform: all members of a species had roughly the same genes and passed these on to their offspring without much change. Clearly, some inherited variation occurs; we see this in bridled murres, black rabbits, pin versus thrum primroses, and mammals with different blood types. But in the past, biologists believed that the proportion of variable gene loci—that is, genes that have slightly different forms in a population (technically called "alleles")—was thought to be very small. Indeed a simple calculation apparently showed that too much genetic variation could not be tolerated: it produced a *genetic load* that reduced fitness and crippled the reproductive potential of the population.

Today we know that this assumption of little variation—near genetic homogeneity—is too simplistic. It has had to be revisited because of experimental results. In 1966 Harry Harris working in London (on human material), and Jack Hubby and Richard Lewontin in Chicago (working on the fruit fly *Drosophila pseudoobscura*) demonstrated that considerable genetic diversity existed. Such diversity can be conveniently measured by heterozygosity, the proportion of genes in an individual where different alleles of a gene are inherited from its parents. Many studies have shown that heterozygosities of 10 percent or more are commonly found in a wide range of organisms (see figure 10).

FIGURE 10. Percentage of heterozygous genes per individual (i.e., different forms [alleles] inherited from each parent) as measured by protein electrophoresis.

Drosophila		15.0
Other insects		15.1
Land snails		15.0
Marine snails		8.3
INVERTEBRATES		14.6
Fish		7.8
Amphibians		8.2
Reptiles		4.7
Birds		5.4
Mammals:	Rodents	5.4
	Large mammals	3.7
VERTEBRATES		5.0
PLANTS		17.0

Different classes of genes have different levels of heterozygosity, but all show high levels.

Two consequences of this high heterozygosity are extremely important in understanding how an island species may differentiate from its originating population. First, a small group of individuals drawn from a large population will almost certainly differ from its parental group in the frequency of alleles at a large number of genes. Second, some alleles are likely to be either absent or relatively overrepresented in the descendant group. This small-group effect will be particularly important if only a small band of organisms manages to colonize an empty island or habitat. This daughter group will be immediately different from its source population, which is, of course, its parents. These effects only became evident after the time that MacArthur and Wilson were developing their ideas and probably explains why they erred in their discussion of the origin of island species.

MacArthur and Wilson were not particularly interested in the processes of speciation. But what they did say about evolution of species takes us to the next stage of the debate about speciation. The animals or plants that colonize an empty habitat constitute the founders of a new population. Their characteristics lead us to consider the *founder effect* or principle. MacArthur and Wilson touched on this question in a chapter on evolutionary changes in their famous book. They begin by saying that since "we believe that evolution through natural selection has produced the biotic differences which characterize islands, it is appropriate for us to study how natural selection works on islands." They continue:

> We can think of the evolution of the new population as passing through three overlapping phases. First the population is liable to respond to the effects of its initial small size. This change, if it occurs at all, will take place quickly, perhaps only in a few generations. The second phase, which can begin immediately and must continue indefinitely, is an adjustment to the novel features of the invaded environment. The third phase, an occasional outgrowth of the first two, consists of speciation, secondary emigration and radiation.

MacArthur and Wilson explicitly equated their first phase with the founder effect, a concept put forward in 1942 by the Harvard

taxonomist and evolutionist Ernst Mayr in one of the defining
works of the neo-Darwinian synthesis, but described more fully in
his 1954 essay "Change of Genetic Environment and Evolution."
For their part, MacArthur and Wilson regarded the founder effect
as "an omnipresent possibility but one easily reduced to insignif-
icance by small increases in propagule size, immigration rate, or
selection pressure . . . The founder principle is actually no more
than the observation that a [founding] propagule should contain
fewer genes [alleles] than the entire mother population."

This is wrong, however. It would have been a reasonable assumption before the discovery of high levels of heterozygosity, but the
post-1966 revolution showing the enormous genetic variability in any group of organisms means that the founder effect will
almost certainly change allele frequencies as well as reduce variability to some extent. The geneticist Sewall Wright, one of the
founders of modern population genetics theory, makes this clear.
He pointed out that Mayr used founder theory to explain only the
loss of genetic variability, not its distribution. He referred to Mayr's
emphasis on the founder effect as leading to gene (or allele) loss
and a reduction in variability, but for Wright the "most significant"
genetic feature in the founding of a species was

> the wide random variability of gene frequencies (not fix-
> ation or loss) expected to occur simultaneously in tens of
> thousands of loci at which the leading alleles are nearly
> neutral, leading to unique combinations of gene fre-
> quencies in each of innumerable different local popula-
> tions. . . . The effects attributed to the "founder effect" by
> Mayr [and also by MacArthur and Wilson] are the most
> obvious but the least important of the three.

In other words, when a small population colonizes an empty
habitat like an island, that population will almost inevitably be different from its source population. The effect is that every colonizing

frequencies of allelomorphs at one locus in population

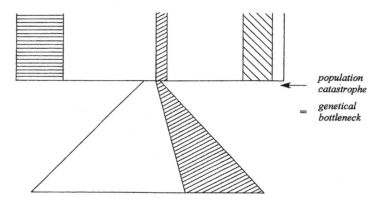

population
catastrophe

= genetical
bottleneck

FIGURE 11. Effects on the frequencies and numbers of alleles at a single gene locus as the result of a bottleneck in numbers of a population. This effect is especially marked when a small number of individuals found a new population. Similar loss of alleles and changes in frequency occur at every locus in the genome, resulting in the founded population being genetically different from the ancestral one.

event will be a novel experiment, exposing to the environment a set of reaction systems determined by the alleles carried by the founding group. It is these systems that represent the new generation's ability to respond (or not to respond, given that most colonizations result in rapid extinction because of the failure of the animals or plants to cope in their new situation) to natural selection. This response is phase 2 in the MacArthur and Wilson scheme; it will necessarily be limited by the chance collection of alleles present in the founders.

This "instant" result of a founding event is obvious once the effect of taking a small number of individuals from a genetically heterogeneous source is realized. Some theoreticians still doubt the importance of founder effects in speciation, but a growing number of studies show their effect in establishing local genetic heterogeneity and hence the possibility of further differentiation. Indeed the founder effect has been described as perhaps the most novel and influential contribution of the twentieth century to ideas about how evolution occurs on islands. There seems no reason to doubt that it is the main determinant of the distinct island races.

It is probably helpful to distinguish the *founder principle*, which occurs through the chance collection of genes carried by founders, from *founder selection*, which occurs following isolation; the founder effect is the overall result of both processes. Separating the founder principle from founder selection is usually impossible because we rarely know the frequencies of genes carried by the founding group. However, reconstructing the genetic constitution of the founders in some human isolates has been possible.

One example comes from the human population of Tristan da Cunha, a volcanic island in the South Atlantic Ocean. The island has the highest frequency in the world of a recessively inherited progressive blindness, *retinitis pigmentosa*. Fifteen individuals effectively founded the Tristan population. We know the pedigree of them all from the time the community was established in 1817 (when the military garrison intended to prevent the escape of Napoleon from the sister island of St. Helena was withdrawn) to the time the population was evacuated in 1961 following an eruption of the island volcano. One of the original founders of the population must have been carrying an allele of the gene affecting retinitis pigmentosa. In technical language he or she must have been heterozygous for the retinitis pigmentosa mutation. Since we all get a set of chromosome from each of our parents, the frequency of that mutation was at least one in thirty—about 3 percent. The population increased seventeen-fold during its period of isolation, but the frequency of the allele remained constant. There were seventeen copies of the allele in 1961—four homozygotes with two copies each of the gene and nine heterozygotes, each with a single copy.

In another case, founder selection seems to have been acting in addition on a condition called *porphyria variegata*, an inherited defect of porphyrin, part of the hemoglobin molecule. The porphyria allele causes its carriers to be extremely sensitive to sunburn, so its occurrence is easily recognized. In South Africa, it is carried by about eight thousand individuals, three in every thousand of the descendants of the original white population, although it is very rare outside the country. All the present-day sufferers in South Africa are in thirty-two family groups, each of which can all trace their ancestry back to one of the original forty pairs of white settlers who arrived from Holland in 1688. One million members of the current population have the same surnames as the original forty—implying an increase of 12,500 times in three centuries. However, the porphyria allele is present in only two-thirds of these. Presumably its deleterious effects (particularly in the sunny climate of South Africa) have led to selection against porphyria carriers whether through leaving South Africa, dying from the condition, or failing to have children.

The Case of the Spittle-Bug

One of the best examples of selection taking place after the original founder event is in a series of experiments involving the frog-hopper or spittle-bug, *Philaenus spumarius*, on small islands in the Baltic Sea off the coast of Finland. The spotting pattern on the wing cases (*elytra*) of these bugs distinguishes a range of forms (or *polymorphisms*), each determined by an allele at a single gene locus. On the mainland of Finland, one form of spotting pattern on the wings is more common toward the north, while another form has a higher frequency in more humid areas, implying that the insect wings are adapting to different conditions.

The island races of the spittle-bug tend to reflect the variability found in mainland populations. Although several of the wing casing (elytral) forms are missing on the outer islands, the same forms also disappear at the northern edge of the species range on the mainland, despite the fact that the bugs there are in breeding contact with a large southern population. In other words, the genetic structure of the island races seems to be based on their adaptation to available plant life, not merely the result of more of the bugs arriving or leaving the areas.

Olli Halkka, a Finnish biologist, and his colleagues found that

some forms of the bug seemed to be particularly hardy and survived better than others when they were introduced to islands
where they did not occur. But the survival advantage altered as
the vegetation changed. The original colonizers were replaced by
another form that had a preference for feeding on newly dominant
plants, like purple loose-strife (*Lythrum salicaria*) and sea may-
weed (*Triplospermum maritimum*).

Direct proof of selection on the islands came from an experiment in which approximately eight thousand individuals (three-
quarters of the populations) were swapped between two islands
with genetically different populations. After three generations, both
island populations had reverted to the pretransfer allele frequencies, although they deviated considerably from another thirty-five
island populations in the area. In other words, the genetic makeup
of the island populations was not random (i.e., determined by
genes introduced by the founding animals) and must therefore be
regulated by natural selection. These Finnish island spittle-bugs are
one of the few cases where separating the genetic founder effect
from subsequent selection has been possible.

JOSEPH HOOKER AND ISLANDS

At this point, we return to Joseph Hooker. Islands brought Hooker
and Darwin together. Darwin wrote to Hooker soon after the latter disembarked from his four-year round-the-world voyage on the
Erebus:

> I had hoped before this time to have had the pleasure
> of seeing you & congratulating you on your safe return
> from your long & glorious voyage. I am anxious to know
> what you intend doing with all your materials—I had
> so much pleasure in reading parts of some of your let-
> ters [sent to Charles Lyell, Senior, father of the well-
> known geologist], that I shall be very sorry if I, as one of

the Public, have no opportunity of reading a good deal more. . . . Henslow [Darwin's botanical teacher] (as he informed me a few days since by letter) has sent to you my small collection of plants—You cannot think how much pleased I am, as I feared they wd have been all lost. I paid particular attention to the Alpine flowers of Tierra Del [Fuego] & I am sure I got every plant which was in flower in Patagonia at the seasons when we were there.— I have long thought that some general sketch of the Flora of that point of land, stretching so far into the southern seas, would be very curious.

Hooker responded quickly:

I am exceedingly glad to think you attach so much importance to the comparison of the Arctic plants with the Antarctic as it was my aim throughout to establish an Analogy between the two hemispheres, & to draw up tables upon several plans, shewing for instance the proportion of plants in each of the predominant Nat[ural] Ord[er]s. common to both. . . . In my Antarctic flora I intend (following my fathers advice) to include Ld Aucklands & Campbells Islds as they contain the most southern plants of those longitudes, & as they have all the nameless peculiarities of plants of high latitudes, quite as much so as those of Fuegia (however luxuriant the vegetation may be compared with analogous Northern latitudes). . . . The Vegetation of Kerguelens Land is entirely that of Southernmost America, almost all its plants being common to the two, few in proportion common to it & Ld Aucklands & none peculiar to the two latter (perhaps one is). The Falkland Isld. flora seems to combine the Patagonian with the Fuegian, I think of including it with the latter.

Darwin and Hooker remained friends and mutual critics. Hooker was the first person told about Darwin's evolutionary ideas. He spoke at the infamous debate in Oxford in 1860 between T. H. Huxley and the bishop of Oxford. Hooker also set out definitive principles about island biotas during the 1866 meeting of the British Association for the Advancement of Science. His lecture was historically significant in the scientific support Hooker gave for Darwin's *Origin of Species*, but has continuing relevance as a penetrating analysis of an important evolutionary situation. The lecture was the first systematic statement of the importance of islands for evolutionary studies. Hooker's identification of the main characteristics of island biotas still stands:

- ▸ They contain a high proportion of forms found nowhere else (*endemics*), although these endemics are usually similar to those found on the nearest continental mass.
- ▸ They are impoverished in comparison with comparable continental areas, that is, there are fewer species on islands than on mainlands.
- ▸ Dispersal must play a part in the colonization and establishment of islands, unless the island has been cut off from a neighboring area and therefore carries a relict of a former continuous fauna and flora.
- ▸ The relative proportions of different taxonomic groups on islands tends to be different from non-island biotas—that is, there is taxonomic "dysharmony."

Peter Grant, who through his careful and long-continued studies on "Darwin's finches" on the Galapagos has probably done more than anyone else to elucidate evolution on islands, has written, "An outstanding feature of islands is their strangeness; many of them are downright weird. Naturalists of the last three centuries brought back to civilization accounts of strange and unimagined creatures found only on remote islands. Dodos. Sphenodon. The Komodo dragon. Daisies as tall as trees. What is it about islands that promotes such strangeness?" Hooker did not know the answer to this, but he was a pioneer in beginning the search, and his work inspired—and continues to inspire—his successors. Through his pioneering insights, we can now start to answer reasonably Grant's question about the curiosities of so many of the plants and animals on islands.

Notes

1. The official naturalist was Robert McCormick, who left the *Beagle* early in the voyage in pique at Darwin's position.
2. r-strategists are opportunists—quick to establish and reproduce themselves, but poor at competing with other species; K-strategists are good competitors, usually with a larger body size, slower development rate, and a lower reproductive capacity than r-strategists. K-strategists dominate established communities and have a lower rate of species extinction than the more ephemeral communities characterized by r-strategists.

Malcolm Jeeves and Warren S. Brown
The Human Animal

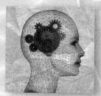 THE 1990s "Decade of the Brain" in science has passed, taking us to the era when the human "mind" is the new frontier, according to Jeeves and Brown in their book, *Neuroscience, Psychology, and Religion*. Human beings share much with other animals. Yet the human mind stands apart in its production, for example, of language and symbolic reasoning. Comparing humans with other animals creates no problem for religious belief, according to Jeeves and Brown. The "war between science and religion" only arises when people insist that humans are "nothing but" animals, or nothing but spirit or soul. The authors argue for a partnership between science and religion. This is possible now that the "cognitive revolution" in psychology has replaced the older Freudian and behaviorist viewpoints, which were hostile to religion. The cognitive revolution includes the field of evolutionary psychology, which studies the origins of the brain, behavior, language, and perception. As in the past, brain science continues to look for "spots" in the brain that produce human experiences, and hence the

argument for a "God spot," or place that produces religion. The authors argue for a more complex and holistic
view of the brain. They find the model of "emergence"—
that the whole is larger than the sum of its parts—best
explains unique qualities to the human mind.

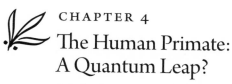

CHAPTER 4

The Human Primate:
A Quantum Leap?
Malcolm Jeeves and Warren S. Brown

IT IS HARD to avoid popular headlines today about the new "evolutionary psychology," which seemingly purports to explain the evolutionary origins of every kind of human behavior. Such popular reports—on everything from courtship between men and women to our responses to marketing—splash across the covers of *Time* and *Der Spiegel* magazines, not to mention scientific journals such as *Science* and *Nature*. This approach has also gained notoriety, and stirred controversy, under the name "sociobiology."

The appeal relates to how evolution tries to show how human beings, on the one hand, are similar to animals (and even ant colonies!) but, on the other, are somehow unique—we emerged from other species with a whole range of specialized and "superior" abilities. Our relationship to the various species of nonhuman primates brings this topic closest to home. This discussion has major implications for both humanist and religious views about human nature. Overall, as research advances, lines that formerly had seemed to divide animals and humans now become increasingly blurred, making the question of human uniqueness ever more problematic.

What is today called "evolutionary psychology" used to be labeled "comparative psychology." It has an illustrious history in psychology's development. An important assumption in comparative psychology was voiced at the beginning of the twentieth century by Lloyd Morgan, the British scientist. Morgan's famous canon proposed that "in no case may we interpret an action as the outcome of a higher psychical faculty if it can be interpreted as the outcome of one which stands lower in the psychological scale."[1]

Early comparative psychology focused on finding links between changes in sensory processes and learning as one moves up the phylogenetic tree of evolution. This, in turn, led to attempts to link the increasing complexity of the brain and central nervous system with more elaborate behaviors and learning capacities. Soon enough, the findings of ethologists such as Conrad Lorenz and Nikolaas Tinbergen revealed the fallacy of believing that increased complexity of the nervous system always meant increasing complexity of learning capacity and social behavior. Some of the animals with the simplest nervous systems showed remarkably complex forms of social behavior—consider, for example, the ants and the bees.

Evolutionary psychology bases its study of human thinking and behavior on the evolutionary principles of natural selection. Accordingly, it presumes that natural selection favored genes that engendered our ancestor with behaviors and brain-processing systems that solved survival problems, thus contributing to the spread of their genes. In 1992 John Tooby and Leda Cosmides defined evolutionary psychology as "psychology informed by the fact that the inherited structure of the human mind is a product of evolutionary processes."[2] Thus, the main focus of research in evolutionary psychology is the question of how humans came to be the special animal that we seem to be.

According to evolutionary psychologist Richard Byrne, some of the central questions in this field are: When did a particular cognitive trait enter the human lineage? What was the trait's original adaptive function? Has it been retained for the same reason, or is it now valuable for some different purpose? Furthermore, what is the cognitive basis for the behavioral trait, and how does its organization relate to other mental capacities?[3]

In discussions of comparative and evolutionary psychology, anthropomorphism figures large. As a result, popular reporting easily gives the impression that it is unproblematic to interpret the behavior of animals closest to us on the evolutionary scale based on our human experience. Frans de Waal comments:

> This use of anthropomorphism as a means to get at the truth, rather than an end in itself, sets its use in science apart from use by the lay person. . . . The ultimate goal of the scientist is emphatically not to arrive at the most satisfactory projection of human feelings onto the animal, but rather at testable ideas and replicable observations. Thus, anthropomorphism serves the same exploratory function as that of intuition in all science, from mathematics to medicine.[4]

To illustrate some of the hottest research questions in contemporary evolutionary psychology, we shall review four research topics: (1) language, (2) "theory of mind" and mirror-neurons, (3) social intelligence, and (4) altruistic-looking behavior in animals. These topics also suggest how this research is relevant to wider issues of understanding the nature of human uniqueness and how these issues prompt a reexamination of some of our religious views of our human nature.

LANGUAGE

Perhaps the greatest evolutionary leap between the mental capacities of the most intelligent nonhuman primates and that of human beings lies in the use of language. However, the apparent size of the chasm traversed has been reduced by the extensive work done over the last twenty years in attempting to teach great apes a language system. The research on ape language has revealed much not previously known about the capacity of apes to communicate using an abstract system of symbols or gestures (including the ability of some to understand human speech). While the linguistic performance of apes has clear limitations, much can be learned about the evolutionary roots of language by studying nonhuman primates.

The natural signaling systems that apes use in the wild are basically closed systems composed of a finite number of basic vocal expressions. These primarily communicate emotional states elicited by a few types of environmental stimuli. Such complexity as exists in the primate's natural signaling system results from gradations and modulations of this small number of basic signal types (not unlike emotional prosody in human language). Thus, apes in the wild cannot be considered to have developed language.

Explicit attempts to teach expanded, abstract communication systems to apes have met with some success. Most of these experiments have involved attempts to teach chimpanzees to communicate with a human being using abstract symbols or tokens or using the gestures of American Sign Language. However, the meaning of this research with respect to a language system (versus rote learning) has been controversial. Apes clearly show the intent of communicating with an abstract system. They have mastered a limited multiword vocabulary and demonstrated an ability to use multiword expressions (two to six words at most) that appear to make sense. Still, we are not sure whether apes instructed in this kind of abstract communication truly possess a structured, rule-governed grammar. Many linguists argue that they have been taught a more sophisticated signaling system, but hardly a language.

Most of the early attempts to teach a language system to chimpanzees began with adult animals. More recent work suggests that chimps raised from infancy in a language-rich environment (more like the experience of human children) obtain a higher level of ability. In their work *Kanzi: The Ape at the Brink of the Human Mind* (1994), American primatologists Sue Savage-Rumbaugh and Roger Levin describe the remarkable language capacity of a bonobo (or pygmy chimpanzee) named Kanzi. As an infant, Kanzi was a passive participant in unsuccessful attempts to teach language to his mother. When finally allowed to express himself via the language system that was being taught to his mother, Kanzi seemed to know spontaneously how to communicate with the symbols and to have developed an unusual (for a chimpanzee) general language processing capacity. Most remarkable was Kanzi's grasp of spoken English. Kanzi eventually was capable of understanding a wide variety of spoken sentence types (thirteen in all), including sentences with embedded phrases. Kanzi responded correctly on 74 percent of 660 novel sentences, including some recognition of semantics carried by word order. This capacity was considered roughly comparable to that of a normal two-and-a-half-year-old human.

In 2007 a report from a meeting on "The Mind of the Chimpanzee" suggested that there is now strong evidence that language started with gesturing. Amy Pollick and Frans de Waal carefully studied four groups of apes living in captivity. Two of the groups were chimpanzees and two were bonobos (pygmy chimpanzees). The researchers videotaped the groups' behavior for hundreds of hours over more than a year. They concentrated on recording facial and vocal expressions, hand and foot gestures, and the behavioral context in which these expressions and gestures took place.[5]

Their hypothesis was that the meaning of facial expressions is hardwired by evolution, whereas the meaning of gestures is learned and, to some extent, arbitrary. The researchers found that facial expressions always occurred in the same contexts in different groups and different species. This was not true for gestures, however. Half of the routine gestures had completely different meanings in the two species. Even within a single group, the meaning of a gesture varied with the context. Gesturing, it seems, is a likely forerunner of language: it comprises more arbitrary links between the physical signal and its meaning. In this context, it is worth remembering that gesture remains a crucial part of human language—just watch anyone walking along using a mobile phone! Evolution does not come up with complicated new structures or capacities in a single leap but builds them up step-by-step. Pollick and de Waal believe that the capacity for speech is built on mental attributes that were acquired millions of years ago when the ancestors of apes and humans began to gesture meaningfully at each other.

Another theory suggests that the key component of language ability is focused in the left hemisphere of the human brain and in a specific gene called protocadhedrinXY.[6] This gene, so it is claimed, can be said to define our species. On this view, the emergence of *Homo sapiens* was not a gradual or continuous process; instead, there is the possibility that, 100,000 to 150,000 years ago, a relatively abrupt jump gave rise to our species. The jury is certainly still out on judging this recent proposal.

Impressed as we are by Kanzi, such outcomes do not emerge spontaneously. They depend on support from a human linguistic community. Unlike Kanzi, apes in the wild develop no more than a contextually and emotionally modulated set of vocal signals, and they acquire some communicative gestures. It takes human instructors to teach adolescent or adult apes to use symbols or gestures with anything like linguistic properties. But again, the fact of Kanzi is an amazing finding: young chimpanzees extensively exposed to human language can understand it in ways not measurably different from a normal two-and-a-half-year-old human. After that point, of course, young human children (three years old) move rapidly ahead with a far more sophisticated expressive grammar and much greater linguistic creativity.

THEORY OF MIND

For any species, dealing with other members is a challenge for which evolution has produced a number of specialized abilities. In primates, a set of highly complex cognitive abilities is the very key to a successful social life. We humans have an irresistible tendency, for example, to translate our understanding of the behavior of others into an assumption about their mental states. We represent what people are doing in terms of what we believe they want and of what we believe they know and do not know. It is this ability that has come to be known in cognitive science as "theory of mind." Because we cannot see what is in a person's mind, we make a reasonable inference (theories) about that mind by what the person does or says.

As far back as 1978 David Premack and Guy Woodruff described animals that had the ability to understand the mind of another, becoming the first to refer to this ability as a "theory of mind."[7] Andrew Whiten and Richard Byrne (two leaders in the field of evolutionary psychology) provide a useful definition:

> Having a theory of mind or being able to mind-read concerns the ability of an individual to respond differentially, according to assumptions about the beliefs and desires of another individual, rather than in direct response to the others' overt behavior.[8]

Comparative and evolutionary psychology both ask the same question: Do nonhuman primates represent to themselves the behavior of other members of their species in a similar mentalistic way? In the absence of language, the task of studying this form of mind reading in animals is difficult. Developmental psychologists, for example, use language to study what children know about the beliefs of others. But in the absence of language, what methods are available to primatologists?

Basic to a theory of mind is the ability, first, to read the focus of another's attention and, second, to comprehend the intentions signaled by the actions of another. These basic forms are amenable to study in primates. Inferring the focus of attention of another has been studied extensively in chimpanzees and is reported in the literature on deception—what is called "Machiavellian intelligence."[9] The capacity of reading another individual's intentions begins in children at age two-and-a-half, and is fairly well developed by age four. As far as we know, a primate's ability to read intentions never gets to the level of the four-year-old human. Similarly, there is no evidence in nonhuman primates of understanding false beliefs—the litmus test for the presence of a theory of mind in human children.

Most of the evidence for a theory of mind in primates has come from observational studies and only recently (and to a limited extent) from experiments. Byrne and Whiten, in field observations of primates, described in their two volumes on Machiavellian intelligence frequently witnessed behaviors of the following kind: They see a female baboon interested in a young male; however, the dominant male would promptly intervene if he saw them doing things like grooming each other, the sort of exchanges that forge bonds among baboons. So, on this particular occasion, a deceptive rendezvous seems to unfold.

It begins when the young male moves behind a rock so he is not visible to the dominant male. The female moves herself slowly but steadily toward the rock, seemingly just looking at the details of the grass and the ground. Eventually, she moves behind the rock, but she does this in a way that the upper part of her body can still be seen from the other side. At the same time, she starts grooming the completely hidden young male, who remains invisible to the dominant male. So the dominant male can't see the offending action, but he can see where the female is, and the female, at the same time, can watch the movements of the dominant male. With respect to a theory of mind, the important questions are: Do the young male and female understand how clever they are being? Are they acting "according to a plan" to conceal their offending actions? Given that the female makes herself visible, is she somehow reckoning that the dominant male will mistakenly think she is alone behind the rock?

Further links of evolutionary psychology with social neuroscience have been made by those who argue that the capacity for certain forms of deception in some nonhuman primates (which looks like mind reading in humans) has resulted from a rapid evolutionary increase in neocortical volume. They point out that there is a direct relationship between neocortical volume and amount of such "clever-looking" behaviors. This relationship with neocortical volume also applies to deception, innovation, and tool use. Richard Byrne has written:

> Quite what benefits a large neo-cortex brings—the underlying cognitive basis of monkey and ape social sophistication—is not straightforward to answer. It is tempting, but may be utterly wrong, to assume that an animal that works over many months to build up a friendly relationship has some idea of the effect its behavior is having on the mind of the other. . . . We assume the agent realizes that by producing a false belief in his victim may risk losing a friend or gaining an enemy. The alternative is a more prosaic mixture of genetic predisposition and rapid learning—and often this is more likely.[10]

The views held by evolutionary psychologists can change rapidly. This is illustrated by the research on theory of mind. It used to be argued that only humans have a "theory of mind." Up until 2000 Michael Tomasello believed that the observational data reported by Byrne and Whiten claiming to show forms of rudimentary "mind reading" in chimpanzees were not convincing evidence that nonhuman primates have a theory of mind.[11] Today, Tomasello, on the basis of his own laboratory studies, is convinced that his earlier views are wrong and in need of revision. In 2003 he wrote:

> In our 1997 book *Primate Cognition* we reviewed all the available evidence and concluded that nonhuman primates understand much about behavior of conspecifics but nothing about their psychological states, [but] . . . in the last five years new data have emerged that require modification of this hypothesis. The form that a new hypothesis should take is not entirely clear, but we are now convinced that at least some nonhuman primates—

the research is mainly on chimpanzees—do understand at least some psychological states in others. . . . For the moment we feel safe in asserting that chimpanzees can understand some psychological states in others, the question is only which ones and to what extent.[12]

The developmental psychologist Simon Baron-Cohen has suggested that autism demonstrates what human life would be like without a "theory of mind." He says that the kind of deceptive behavior documented in nonhuman primates is no trivial achievement. The deceiver needs to have the mental equipment to juggle different representations of reality. He further notes that the neurological condition that leads to difficulties in socializing, in chatting with others, also leads to difficulties in recognizing when someone might be deceiving them, and that nicely sums up some of the major problems faced by people with autism. Baron-Cohen comments:

> Many children with autism are perplexed by why someone would even want to deceive others, why someone would think about fiction or pretence. They have no difficulty with fact (what he calls version 1 of reality) and can tell you easily if something is true or false ("Is the moon made of rocks? Yes! Is the moon made of cheese? No!"). . . . They may be puzzled by version 2 of reality that, "John believes the moon is made of cheese." Why should a person believe something that is untrue?[13]

In this sense, Cohen believes such children show some degree of "mindblindness." They do not have a fully developed theory of mind. As he explains further:

> Even the higher functioning children on the autistic spectrum, such as those with Asperger's syndrome,

show delays in the development of mind-reading ability. This neurological (and ultimately genetic) set of conditions can leave the person with autism or Asperger's syndrome prey to deception and exploitation.[14]

In a word, this new kind of research in evolutionary psychology may provide medical insights into some extremely trying neurological conditions that up to now have been very difficult to fathom.

Mind Reading: The Mirror Neuron Story

If an actual mechanism for the so-called mind-reading behavior of primates could be found, it would be a natural bridge between neuroscience, evolutionary psychology, and social cognition. Some have heralded such a discovery in a neural substrate, or cell tissue, called the "mirror neuron."

The mirror neuron story began twenty-five years ago when Italian neurophysiologist Giacomo Rizzolatti and his colleagues reported the discovery of neurons in the frontal lobes of the brains of monkeys, which possessed functional properties not previously observed.[15] The unusual property of these cells was that they were active not only when a monkey initiated a particular action but also when the animal observed another monkey initiating and carrying out the same action. For this reason, they were labeled by some as the "monkey-see-monkey-do" cells. These unusual neurons did not respond when the monkey was merely presented with a conventional visual stimulus. Rather, they were activated only when the monkey saw another individual (whether the human experimenter or another monkey) making a goal-directed action with a hand or mouth.

Vittorio Gallese (a collaborator with Rizzolatti in the study of mirror neurons) speculated that a primary and important role of mirror neurons is that they underlie the process of "mind reading," or are at least a precursor to such a process.[16] The potential importance of this work on mirror neurons for understanding social cognition was also recognized by V. S. Ramachandran (a research neurologist at the University of California, San Diego). He wrote, "I predict that mirror neurons will do for psychology what DNA did for biology"—a truly bold and far-reaching suggestion.[17] Ramachandran believes that understanding the role of these cells may give us a deeper insight into how we assign intentions and beliefs to other inhabitants of our social world. Thus, the discovery and description of the responses of these important neurons represent the convergence of state-of-the-art research in neuroscience, evolutionary biology, and psychology.

Functional brain imaging has made it possible to examine the neural substrates of movement production, perception of action, and imagery in humans. Results indicate that witnessing the hand movements of another person activates the prefrontal cortex in the homologous area where mirror neurons were found in the monkey. Thus, activity in this area of the human brain is believed to constitute the neural basis for imitating the actions of others, as well as inferring the intentions of those actions (that is to say, "mind reading"). As you might expect, there is great debate among active workers in this field about what constitutes true imitation, what constitutes mind reading, and what might be the relationship between data from brain imaging in humans and direct nerve-cell recordings in primates.

SOCIAL INTELLIGENCE

In their 1988 book *Machiavellian Intelligence*, Byrne and Whiten looked at the complexities of the social life of our ancestors as a possible route to understanding the development of our distinctive abilities. In 1997, in a second volume under the same title, they extended their findings and argued that intelligence began in social manipulation, deceit, and cunning cooperation. Humankind is, of course, at the apex of this evolutionary development of social intelligence. Thus, Whiten further argues that what differentiates human society from chimpanzee society is the level of cognitive sophistication at which social integration and interaction occurs. Whiten calls this distinguishing feature a "deep social mind" and further claims that "humans are more social—more deeply social—than any other species on earth, our closest primate relatives not excepted. . . . By 'deep' I am referring to a special degree of cognitive and mental penetration between individuals."[18]

Careful observations in the wild and detailed testing in captivity have both produced a wealth of new data on chimpanzee social behavior—so much so, in fact, that Whiten recalls that, fifty years ago, when so little was known, the current findings about a complex chimpanzee society could have not been imagined. This view is endorsed by de Waal, who points out that, even a decade ago, there was no firm consensus on chimpanzee society. Today, there is little debate. Nevertheless, in evaluating the evidence for sophisticated social behavior in animals, we should follow Byrne's warning of the dangers of drawing inferences from field observations alone:

> Researchers have to be very cautious, then, in attributing to non-human primates the ability to understand social behavior or how things work in the mechanistic way of adult humans. Rapid learning in social circumstances, a good memory for individuals and their different characteristics, and some simple genetic tendencies are capable of explaining much that has impressed observers as intelligent in simian primates.[19]

If humans have a "deep social mind" compared to a chimpanzee, what is it in brain structure that might account for this difference? Recent research has proposed that the critical element is an enhancement in neural wiring—namely, a proliferation of Von Economo neurons. These neurons are very large neurons that have very long axons projecting throughout much of the cerebral cortex. The origin of Von Economo neurons is the limbic cortex

—specifically, the anterior cingulate gyrus and insula of the cortex. The insular cortex receives information about the state of the body (that is, visceral/autonomic information, including feedback from bodily responses that would occur in emotions). The anterior cingulate cortex consistently shows activity when an individual is making decisions in the social or moral domain and during the experience of social emotions. Under neuroimaging, both the anterior cingulate cortex and the insula showed marked activity during states of empathy, shame, trust, and humor, as well as during detection of the mental and emotional states and during moral decision making.

According to the theory of Caltech neuroscientists John Allman, Patrick Hof, and their colleagues, the experience of bodily emotions converges on the anterior cingulate and insular cortex, which spread this information through the cortex by way of the Von Economo neurons. This process informs the cognitive functions of these emotional states.[20] This final integration of bodily states with higher cognition allows the human brain to comprehend emotion itself, thus signaling to a person the social significance of actions and perceptions.

The Von Economo neurons are important in any discussion about the "deep social mind" of humanity because they are relatively unique to the human brain. This type of neuron is found in great abundance in the adult human brain and in the brain of a four-year-old child, but they are few in number in newborn human infants and in apes and entirely nonexistent in lower primates.[21] It is also of interest that these neurons have been found to be about 30 percent more numerous in the right hemisphere of the human brain, often thought to be particularly involved in the processing of emotional information.[22]

Evolutionary psychology has tended to reject any theoretical efforts to separate cognitive capacities (like language) from social capacities and experiences because it views these two domains as integrated in mutually reinforcing ways, finally making the complexity of the human being unprecedented compared to monkeys and great apes. Social interactions, therefore, seem to be crucial in shaping human behavior and priming the appearance of the highest and most complex cognitive skills, including language.

ALTRUISTIC BEHAVIOR IN ANIMALS

Over the past three decades, researchers have increasingly observed that some animal behaviors, if seen in humans, might be called moral or altruistic. "Aiding others at a cost or risk to oneself is widespread in the animal world," according to primatologist de Waal.[23] Attempts to understand such "altruistic" behavior had been closely linked to a genecentric sociobiology. According to this view, genes favor their own replication: a gene is successful if it produces a trait that, in turn, promotes the continuance of the gene. To describe this idea of genetic self-promotion, the Oxford zoologist Richard Dawkins introduced the psychological term "selfish" in the title of his book *The Selfish Gene*. What might normally be called a generous act in common language, such as bringing home food, now was interpreted as actually "selfish" from the gene's perspective.

As the original meaning of the words "from the gene's perspective" were forgotten and then discarded, we now hear a constant discussion in biology about how *all* behavior is selfish. Obviously, however, genes have neither a self nor emotions to make them selfish. The phrase is simply a metaphor. Nevertheless, when a metaphor is repeated often enough—even in science—it can assume an aura of literal truth. Dawkins himself had cautioned against going too far with his anthropomorphic talk about a selfish gene, but to little effect. To redress this distortion, de Waal and other biologists are trying to separate the "selfish" metaphor from the actual scientific findings that show remarkable social behaviors in animals.

Once this confusion is cleared up, evolutionary theory can use the genetic makeup of the organism to try to explain the evolution of a capacity of aid to others. That genetic explanation comes in two general ways:

1. Genes favoring altruism can spread in future generations if the costs of the genetically related altruistic behavior to the altruists' personal reproductive success are outweighed by the benefits in reproductive success of the altruists' relatives carrying copies of the same genes. This is termed *kin selection*.

2. Genes favoring altruism could also spread if the altruism is sufficiently reciprocated (*reciprocal altruism*)—that is to say, the benefit bestowed by the altruist on another is received back in kind.

As regards kin selection, examples are widespread in the animal kingdom. Some of its most extreme forms are found, as one might expect, in those odd species where individuals in the colony are unusually highly related to each other. For example, in social insects like bees and ants, the genetic relatedness of workers to each other and to the queen is three-quarters (whereas the maximum found in mammals is one-half, based on relations between a parent and child and between siblings). This is taken to explain why sterile castes of workers evolved in ants and bees; these workers are totally altruistic, spending their whole lives giving food for the "good of the group" (or the good of the queen, who is the only one to reproduce directly). One of the most graphic examples is "honey-pot" worker ants, who do nothing but hang from the ceiling of the ant colony, acting as receptacles or storage jars for honey, which some workers fill them with and which the colony draws on when needed. At the individual ant level, that is self-sacrifice!

Examples of reciprocal altruism appear to be much rarer. Apart from human beings, there are only a handful of examples. A classic example is the vampire bat. Vampire bats are in real danger of starving if they should fail to get their blood meal on a particular evening. However, if this happens, they are fed back in their colony by an unrelated nest mate, to whom they are likely to repay the favor on another night.

We should not assume, however, that, because two behaviors appear to be similar (for example, in animals and humans), therefore, the underlying behavioral mechanisms are similar or identical. In high-tech laboratories, we can now reproduce aspects of human and animal behavior in robots. Still, no one suggests that the underlying mechanisms producing those behaviors are the same. They may share some common features but, when it comes to questions of motivation, conscious awareness, and goal-directed behavior, the two may be miles apart. Likewise, because we can observe self-giving, self-sacrificing behavior in different evolutionary phyla, that tells us nothing about the underlying mechanisms involved. How, for example, could a behavior be "self-giving" if there is no awareness of "self"?

There are some compelling (but anecdotal) examples of self-giving behavior in nonhuman primates. Field-working primatologist Jane Goodall describes unusual chimpanzee behaviors that are not done by all chimpanzees or even by particular chimpanzees routinely. This includes a female helping her mother, even though the mother is unlikely to help the daughter in return or reproduce their genes again. Such anecdotal observations are scientifically problematic, but they certainly are different from the ant cases already mentioned. Here we have an unusual episode in which the female recognized her mother as in need of help and worked out a way to help her.

As we already emphasized, self-giving behaviors in different animals do not tell us the roots of those behaviors. Self-giving may occur, for example, with or without self-awareness. We also have very persuasive arguments that self-giving and self-limiting behavior in organisms developed during a long evolutionary history and eventually emerged in nonhuman primates. That evolutionary viewpoint does not make the behavior any less worthy, nor does it argue that humans are "nothing but" complex primates. Emotions are easily stirred whenever someone notes the similarity between human and nonhuman primate behavior. Some of these behaviors are clearly related. But the idea that humans are "nothing but" glorified apes ignores the distinctiveness of the ethical, moral, and religious aspects of human cognition and behavior.

This warning against reductionism is brought out in de Waal's 1996 book *Good Natured*, which tells us that

> even if animals other than ourselves act in ways tantamount to moral behavior, their behavior does not necessarily rest on deliberations of the kind we engage in. It is hard to believe that animals weigh their own interests against the rights of others, that they develop a view of the greater good of society, or that they feel lifelong guilt about something they should not have done. . . . To communicate intentions and feelings is one thing; and to clarify what is right, and why, and what is wrong, and why, is quite something else. Animals are not moral philosophers.[24]

Regarding that moral sense, he also writes that "the fact that the human moral sense goes so far back in evolutionary history that other species show signs of it, plants morality firmly near to the centre of our much maligned nature." He adds that "humankind's uniqueness is embodied in a suite of features that include ethical behavior and religious beliefs."[25]

Other leading evolutionary psychologists, who have no religious axe to grind, also criticize the excessive enthusiasm for equating nonhuman primates and humans, warning that such exaggeration can jeopardize the scientific work. Richard Byrne, in commenting on a reported trend for human stepfathers to murder their partner's babies under three years old, warned against forcing such findings among animals into assumptions about so-called evolutionary stable strategies among humans:

> These [behaviors] are not carefully thought out by

beasts; and nor are any genes really selfish or altruistic, they are no more than pieces of DNA molecules; nor is an understanding of kinship likely to be remotely similar to our own. [Concepts like murder and altruism] are human applied labels based on the superficial appearance of the actions of individual animals whose behavior is partially governed by genes. . . . Natural selection is a mechanistic process and thus morally neutral; discovering a genetic influence on murder does not condone it. . . . Human social behavior is [by contrast] influenced by our culture and our extensive information transmission by spoken and written language in ways not well described by biology.[26]

HUMAN DISTINCTIVENESS: SEARCH FOR THE QUANTUM LEAP

Such comparisons between humans and other species have not stopped the search for the "quantum leap," or as some say, "phase change," that was necessary to make humans so different from animals. This is a scientific puzzle that will occupy us for many years to come. There are serious scientific issues to be addressed here, and it may be tempting, in a search for human uniqueness, to seize upon a particular human capacity (such as "mind reading") as one way of uniquely separating off humans from nonhumans. At the same time, when similarities between the behavior of humans and some nonhuman primates are identified, there will be the temptation to say that humans are "nothing but" unusually complex primates.

For most people, however, probably the very first obstacle in this scientific discussion is their feeling that it is demeaning, and even offensive, to compare humans directly with apes. They might be given the wise counsel that, in fact, we can gain much self-knowledge in this comparison. Still, that does not always clear away the emotional obstacles. Perhaps the best way to approach the comparison of ourselves and animals is first to ask, "Is there a difference in kind or merely a difference in degree?" In the seventeenth century, the French mathematician Blaise Pascal approached this type of question by discussing it from a theological point of view:

> It is dangerous to show a man too clearly how much alike he is to the beasts without showing him his greatness. It is also dangerous to show him too clearly his greatness without his lowliness. It is still more dangerous to leave him in ignorance of both.[27]

This discussion also can be confused by the term *uniqueness*, which has several dimensions. Animals of each phylum are unique. Each has properties and abilities none others do. Birds can fly and we cannot (at least unaided). For the religious person who might worry about these modern comparisons of humans and animals, it is helpful to know that evolutionary psychology, as a science, has nothing at stake in the religious question of human uniqueness. The science does not draw a theological conclusion, even though individual psychologists, based on their personal beliefs, may probe these theological questions.

As a science, evolutionary psychology simply hopes that the study of animal behavior will help us detect possible beginnings of language and our own ability to possess a theory of mind, and see the seeds of human culture in chimpanzee societies. We can pursue this research carefully and still avoid the temptation to adopt unjustified implications for human behavior that we glean from observing other animals.

One useful analogy for explaining the remarkable cognitive and social gap between nonhuman and human primates is that of a "phase change," a concept used in physics. This kind of change occurs when the same basic materials (in our case, having a basic animal brain system) suddenly or gradually exhibit new properties. In physics, oxygen and hydrogen in appropriate proportions and

under specific conditions become a liquid with different properties from gases. In another example, the physicist and theologian
John Polkinghorne recounts how the seeming irrationality of the
superconductivity state made sense only when it was realized that

> there was a higher rationality than that known in the
> everyday world of Ohm. After more than 50 years of
> theoretical effort, an understanding of current flow in
> metals was found which subsumed both ordinary con-
> duction and superconductivity into a single theory. The
> different behaviors correspond to different regimes,
> characterized by different organizations of the states of
> motion of electrons in the metal. One regime changes
> into the other by phase change (as the physicists call it)
> at the critical temperature.[28]

Despite this obvious gap between the social and mental life
of humans and other animals, evolutionary psychology has produced evidence that animals can indeed think at some level. Such
behaviors would be typically described as rudiments of imagination, inventiveness, and means-end reasoning. In the case of "mind
reading," evolutionary psychology and neuroscience have developed explanations for how this emerged even in animals. As a result
of this research into the thinking powers of the primates, it has
become difficult to draw a clear demarcation between some of the
fundamental mental abilities of nonhuman primates and humans.

This fuzzy boundary between humans and animals should not
really bother a religious outlook, since the important aspects of
human uniqueness are based on theological presuppositions, not
on neurobiological observations. At the same time, humans are
clearly unique by way of their explosive development of learning,
philosophy, literature, music, art, science, and so on. What needs
to be said about animals is only this: they do show reasoning and
thinking abilities. While this ability is rudimentary, it overlaps with

similar abilities in developing human children. At the most, therefore, scientific research has made it difficult to use the marks of thinking or reasoning as the sole anchor to claim that humans are created in the image of God.

Some Christians may be concerned about this narrowing of the gap between ourselves and some nonhuman primates. For our part, we believe that there are no great issues at stake in this research. As the quotes above demonstrate, the careful scholars and workers in this field also are dismayed by the popular exaggerations in the media. A Christian can be enthusiastically open-minded about developments in evolutionary psychology—not gullible, but discerning, and glimpsing fresh pointers to the greatness of the Creator in the wonders of his creation. Neurobiology and evolutionary psychology are areas of science where we may exercise stewardship by engaging in new research on such distressing mental conditions as autism and, in this way, show care and compassion.

The course of creation has been such that the qualities of self-giving and self-limiting behavior, built upon neural substrates, may be traced out, coming to full flower in humankind. This is not to say that such behavior is genetically determined. Its expression increases and multiplies moment by moment, depending on personal choices and arguably on the catalytic effect of living in a self-giving community. Again, constant vigilance is called for to avoid slipping into sloppy thinking that assumes that similarities in overt behavior demonstrate identical mechanisms for those behaviors.

Speaking personally, from within our shared Christian tradition, we do not find it necessary to deny the emergence of elements of altruistic or self-giving behavior in nonhuman primates in order for us to affirm the reality of what is called *agape* love, seen uniquely in the self-giving and self-emptying of Jesus Christ. The self-giving of Christ was unique, and it is by faith that we affirm that the ultimate act of Christ's self-giving, by its nature, sets him and it apart from all others.

Notes

1. See the entry for Lloyd C. Morgan in *The Oxford Companion to the Mind,* ed. Richard L. Gregory (Oxford: Oxford University Press, 1987), 496.
2. Jerome H. Barkow, Leda Cosmides, and John Tooby, eds., *The Adapted Mind: Evolutionary Psychology and the Generation of Culture* (New York: Oxford University Press, 1992), 7.
3. R. W. Byrne, "Evolutionary Psychology and Sociobiology: Prospects and Dangers," in *Human Nature,* ed. M. A. Jeeves (Edinburgh: Royal Society of Edinburgh, 2006), 84–105.
4. Frans de Waal, *Good Natured: The Origin of Right and Wrong in Humans and Other Animals* (Cambridge, MA: Harvard University Press, 1996), 64.
5. Amy S. Pollick and Frans B. de Waal, "Ape Gestures and Language Evolution," *Proceedings of the National Academy of Sciences of the United States of America* 104, no. 19 (2007): 8184–89.
6. T. J. Crow, ed., *The Speciation of Modern* Homo Sapiens (Oxford: Oxford University Press, 2003).
7. D. Premack and G. Woodruff, "Does the Chimpanzee Have a Theory of Mind?" *Behavioral and Brain Sciences* 1, no. 4 (1979): 515–26.
8. Andrew Whiten and Richard Byrne, *Machiavellian Intelligence II: Extensions and Evaluations* (Cambridge: Cambridge University Press, 1997), 150.
9. Richard Byrne and Andrew Whiten, eds., *Machiavellian Intelligence: Social Expertise and the Evolution of Intellect in Monkeys, Apes and Humans* (Oxford: Clarendon Press, 1988).
10. Byrne, "Evolutionary Psychology and Sociobiology," 91.
11. Michael Tomasello, "Primate Cognition: Introduction to the Issue," *Cognitive Science* 24, no. 3 (2000): 357.
12. Michael Tomasello, Josep Call, and Brian Hare, "Chimpanzees Understand Psychological States—The Question Is Which Ones and to What Extent," *Trends in Cognitive Science* 7, no. 4 (2003): 153.
13. Simon Baron-Cohen, "I Cannot Tell a Lie," *In Character* 3, no. 3 (Spring 2007): 55–56.
14. Ibid., 56.
15. Giacomo Rizzolatti et al., "Premotor Cortex and the Recognition of Motor Actions," *Brain Research/Cognitive Brain Research* 3, no. 2 (1996): 131–41.
16. Vittorio Gallese, "Before and Below 'Theory of Mind': Embodied Simulation and the Neural Correlates of Social Cognition," *Philosophical Transactions of the Royal Society of London. Series B, Biological Sciences* 362, no. 1480 (2007): 659–69.
17. V. S. Ramachandran, interview with Tom Stafford, *The Psychologist* 17, no. 11 (November 2004): 636–37.
18. Andrew Whiten, "The Place of 'Deep Social Mind' in the Evolution of Human Nature," in Jeeves, *Human Nature,* 212.
19. Byrne, "Evolutionary Psychology and Sociobiology," 99.
20. John M. Allman et al., "Intuition and Autism: A Possible Role for Von Economo Neurons," *Trends in Cognitive Science* 9, no. 8 (2005): 367–73.
21. John M. Allman et al., "The Anterior Cingulate Cortex: The Evolution of an

Interface between Emotion and Cognition," *Annals of the New York Academy of Science* 935 (2001): 107–17.
22. Esther A. Nimchinsky et al., "A Neuronal Morphologic Type Unique to Humans and Great Apes," *Proceedings of the National Academy of Sciences of the United States of America* 96, no. 9 (1999): 5268–73.
23. De Waal, *Good Natured*, 12.
24. Ibid., 208.
25. Ibid., 218.
26. Byrne, "Evolutionary Psychology and Sociobiology," 96.
27. Blaise Pascal, *Pensées*, trans. Roger Ariew (Indianapolis: Hackett Publishing, 2005), 33.
28. John Polkinghorne, *The Way the World Is* (London: SPCK, 1983), 55.

Denis R. Alexander
The Language of Genetics

 RESEARCH IN GENETICS today proceeds at a "breathless" pace, says Alexander in his book *The Language of Genetics*. The implications for medicine and the study of biological evolution have been astounding. However, the question of "What is a gene?" is "not so straightforward" as it once seemed, Alexander says. The history of genetics is one of twists and turns, and many blind alleys—typical of science—before it arrives at our current knowledge of the DNA code and how it shapes biological life and, to an extent, human personality. The new field of "epigenetics," however, has revised past beliefs by showing how the environment and human behavior can alter the ways in which genes are regulated. Genes are not as deterministic as once believed: they are like an orchestra that can play the same score different ways. This has philosophical implications, says Alexander, who argues, "The human genome endows us with freedom of choice." For theologians, the question remains of how to reconcile a Creator with the apparent "chance" operation of genes in mak-

ing us who we are. For science and policy makers, the topic of genetic "engineering" tops a list of moral debates, especially when it comes to engineering human destinies. Either way, Alexander says, "Theological accounts are not in any kind of rivalry with the scientific accounts, which are fine as far as they go."

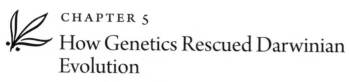

CHAPTER 5

How Genetics Rescued Darwinian Evolution

Denis R. Alexander

FOR MANY YEARS Charles Darwin sought to understand the mechanism of biological inheritance, but he never did find out about genes. Fortunately, failing to understand one biological mechanism did not prevent him from discovering another, and he published his theory of evolution in *On the Origin of Species by Means of Natural Selection* (1859). "Natural selection" is shorthand for differential reproductive success among organisms based on their heritable traits, as explained in greater detail below. It is essentially a simple idea, if not intuitive. After hearing Darwin's new theory, Thomas Henry Huxley exclaimed, "How extremely stupid not to have thought of that!"

Darwin may have felt the same way if someone had been able to explain genetics to him. Surely he would have been delighted when scientists began to do what he had wanted to do but could not—to combine genetics with his theory of natural selection. The resulting "neo-Darwinian synthesis," as it is sometimes called, is what makes contemporary evolutionary theory so powerful.

When scientists use the word "theory," it has a technical meaning different from its everyday usage to express doubt, such as when we say, "Oh well, it's only a theory." A "theory" in science is much more like a map that incorporates and renders coherent a broad array of data. When scientists talk about the "theory of gravity" or the "theory of relativity," they are not expressing any doubt that things in general fall downward, nor that the relative motion of observers changes what they observe. In a similar manner, the theory of evolution acts like a reliable, conceptual map to render coherent a huge array of disparate data, including data derived from genetics, fossils, dating, anatomy, physiology, and the geographical distribution of species.

The aim here is not to present the evidence for evolution, which has been done very effectively in many publications,[1] but rather to focus on the function of genes in the evolutionary process. As the following brief history shows, the human story of finding out how genes work included many twists and turns, reminding us once again that progress in science is far from linear.

THE DECLINE AND REVIVAL OF NATURAL SELECTION

Darwin died in 1882 and was buried in Westminster Abbey with great pomp as a British scientific hero. But ironically, for the following fifty years his theory of natural selection actually declined in popularity, and by 1900 some biologists were talking about the demise of Darwinism. In 1903 the German botanist Eberhard Dennert proclaimed, "We are now standing by the death-bed of Darwinism, and making ready to send the friends of the patient a little money to ensure a decent burial of the remains."

Evolution as an idea remained immensely widespread and popular, but *how* evolution actually happened was widely disputed. A large part of that answer, of course, would be found in the rediscovery of the principle of inheritance first understood by the Moravian monk Gregor Mendel (1822–1884). Having worked in relative obscurity, Mendel and his findings remained unknown to the broader scientific community until around 1900. Mendel showed that genes operate as particles that combine to produce dominant and recessive traits in offspring—the basis of Mendelian genetics. In the absence of Mendel's insights, many biologists adhered to Lamarckian evolution, which had been proposed by the French biologist Jean-Baptiste Lamarck (1744–1829). In this view, organisms evolved by "acquired characteristics." These characteristics were given by the blending of parental traits, by the environment, or produced simply by physical effort, such as generations of giraffes stretching their necks to reach leaves in tall trees. Lamarckian evolution retained great appeal because it also seemed to explain the sudden jumps that were observed in the fossil record, then far more incomplete than it is today.

Even great enthusiasts for evolution, such as Darwin's friend Thomas Henry Huxley, never really accepted slow-acting, incremental natural selection as the mechanism for evolution, much preferring the idea of big jumps or so-called saltations. The secular Huxley was also suspicious of the role of chance in generating variant phenotypes of organisms upon which natural selection then acted. For Huxley, chance sounded like an opening for God's special creation, whereas he wanted to see evolution emerging out of natural scientific laws. It is ironic that in his day Huxley resisted the idea of chance because he thought that it had theological overtones, whereas creationists today resist the idea of chance because they think that it has atheistic overtones. People often interpret essentially the same data in quite different ways depending on their political, economic, and cultural contexts.

The great Victorian idea of progress also seemed to fit better with Lamarckian ideas. Surely it is more rational, so the argument went, that the useful things that animals learn during their lifetimes should be passed on to their offspring. Why waste what you've learned? Let it benefit a future generation. This again illustrates the influence of one's own political or social ideologies upon the interpretation of data. Letting the data speak for itself is not as straightforward as it may seem.

Mendel's results were rediscovered and extended around 1900. You might have imagined that once the Mendelian laws of inheritance had been rediscovered, then obviously they would be brought together with the idea of natural selection to generate very quickly the kind of theory of evolution that we have today. But that didn't happen right away. During the early decades of the twentieth century, Mendelism, as the pattern of inheritance that Mendel discovered became known, was actually seen as a *rival* to the theory of natural selection. How come?

The answer is that the particulate concept of inheritance readily lent itself to the notion that changes in evolution happen rather suddenly. For example, the botanist Hugo de Vries made extensive studies of the evening primrose, observing that new, differently colored varieties sprouted seemingly at random. His so-called mutation theory became the most popular theory of evolution in the early decades of the twentieth century. At that time, the term "mutation" primarily referred to the apparently sudden appearance of a distinct form. Only much later did the word take on its contemporary meaning: a physical change in DNA that may cause a change in phenotype.

The early twentieth century therefore saw Mendel's results on particulate inheritance identified with the idea of mutations, perceived as saltations (sudden jumps), such that speciation itself was thought to happen abruptly. To some biologists this made Darwinian natural selection appear superfluous. If new varieties or mutations could come about suddenly, then why did you really need natural selection? Other biologists, and de Vries was one of them, retained a negative role for natural selection in eliminating the unfit mutational varieties that arose, but they didn't credit natural selection with the power to establish new varieties through incremental adaptation. Instead it was thought that species occasionally went through rapid bouts of mutation in which they generated a whole selection of new varieties, which also explained, so it was thought, the gaps in the fossil record.

Now what all this shows is that it's not a good idea to base general conclusions in biology on the study of just one or a few species. By 1920 it became clear that the evening primrose that de Vries had been studying for so long was a complex hybrid, and his apparently new forms of primrose were not new examples of mutation at all, but simply recombinations of existing characteristics.

New insights came from investigations on quite a different organism. Thomas Hunt Morgan's groundbreaking work on *Drosophila* during the first few decades of the twentieth century showed that genes were strung out on chromosomes "like beads on a string." Morgan was a laboratory-based experimentalist who initially saw little role for natural selection and was greatly impressed by de Vries' mutations. He decided to see if he could produce the same type of saltationist mutations in his flies, but what he found were small but definite variations that were inherited in a Mendelian fashion. Through these observations and many conversations with his students and colleagues, Morgan finally came to accept that Darwin's theory of natural selection could account for the origin of species, although he always emphasized the role of mutation in providing the material upon which selection acts.

BIRTH OF THE NEO-DARWINIAN SYNTHESIS

The next important stage in the development of evolutionary ideas in biology came not from plant breeders, nor from fly breeders—indeed not from the laboratory at all—but from population geneticists and mathematicians. The key question now was, how did evolution actually work in populations of living organisms out in the wild? Three famous figures were associated with this shift in thinking: the mystic British communist J. B. S. Haldane, the Anglican British eugenicist R. A. Fisher, and the American Sewall Wright, the son of first cousins, who was a professor at the University of Chicago.

These three scientists were the first to apply mathematical analysis to the study of genetic variation in a population, and it was the fusion of mathematical genetics with the theory of natural selection that later came to be known as the neo-Darwinian synthesis.

Biologists at that time were so unused to mathematical treatments of their subject that Fisher's first paper submitted to the journal of the London Royal Society was turned down because no one could understand it! However, once explained, it was clear that this new approach was very useful. Most notably, the trio used it to identify four main evolutionary factors that affect how gene frequencies in a population change over time: genetic drift, natural selection, mutation, and gene flow.

Before we look at these four factors, we should elaborate a bit more on the definition of a "gene." By definition, a gene is a unit segment of DNA sequence with an overall function. However, not all copies of a particular gene are identical. So it's necessary to distinguish among the different variants of a gene when examining how these evolutionary forces shape gene frequencies in a population. To understand this process it is useful to know a bit of jargon often used in genetics: the term "allele." An "allele" refers to different versions of the same gene that may be identical or slightly different. For example, healthy and disease-related alleles of the same gene can have very different effects on an organism. Strictly speaking, an allele cannot be defined as "beneficial" or "deleterious" as if this were its permanent characteristic. It all depends. With this understanding of an "allele" in hand, we are now better placed to appreciate the four genetic factors involved in the evolutionary process.

First, if there's no obvious advantage or disadvantage among the alleles of a given gene, the evolutionary factor most likely to describe their frequencies is *genetic drift*. Genetic drift is the change in the relative frequency at which an allele occurs in a population due to random sampling and chance. The alleles in offspring are a random sample of those in the parents, and chance has a role in determining whether a given individual survives and reproduces.

As figure 12 illustrates, imagine that you put twenty marbles in a jar to represent twenty organisms in a population. Half of them are black and half white, corresponding to two different gene alleles in the population. The offspring they reproduce for the next generation are represented in another jar. In each new generation the organisms reproduce at random. To represent this reproduction, randomly select any marble from the original jar and deposit it in the second jar. Repeat the process until there are twenty new marbles in the second jar. The second jar then contains a second generation of "offspring," twenty black or white marbles. Unless the second jar contains exactly ten white and ten black marbles there will have been a purely random shift in the allele frequencies, which will then influence the allele frequencies of the next generation as well. In due course all of the marbles will be entirely white or entirely black, and the other allele will then have been lost.

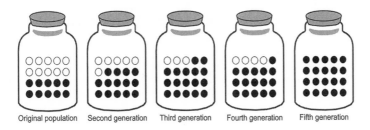

Original population   Second generation   Third generation   Fourth generation   Fifth generation

FIGURE 12. An illustration of genetic drift. The first jar on the left contains a "marble population" in which two colors of marbles represent two different genes at a 50:50 ratio. In each generation ten marbles are selected at random from the jar representing the previous generation. Given enough "generations," one type of marble will always go to fixation and the other will be lost.

While drift constantly and randomly reduces genetic variation, natural selection provides another, more selective winnowing force. It acts as a powerful sieve, just as Darwin had always maintained, filtering out those sets of alleles that reduce the fitness of the organism. Fitness here doesn't refer to the consequences of going to the gym regularly, but is rather a shorthand way of expressing reproductive success.[2] Organisms well-fitted to their environment are those that generate plenty of progeny in succeeding generations. Sewall Wright pictured selection acting on an "adaptive landscape" in which mountain peaks represent well-adapted sets of genetic variants, or genomes, and valleys represent poorly adapted genomes (figure 13).

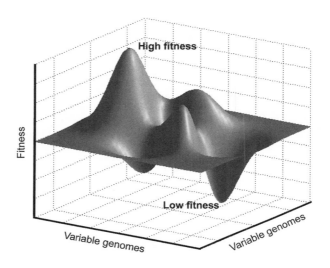

FIGURE 13. The adaptive landscape, suggested by Sewall Wright, is a way to visualize the relationship between genes and evolutionary fitness. Peaks on the landscape represent high fitness while valleys represent low fitness. Different combinations of genes yield different degrees of fitness.

If drift and natural selection were the only forces responsible for shaping the genetic variation of a population, variation would largely disappear. Mutation and gene flow thus serve to add genetic variation to a population. In the context of population genetics, "mutation" in the 1920s and 1930s came to mean a discrete change in a gene, resulting in a distinct allele, even though chemically these changes weren't understood at the time because DNA wasn't yet understood.

Gene flow simply refers to the transfer of alleles of genes from one population to another in the same species. Let's imagine that two animal populations have been breeding quite separately on either side of the country. During their time apart they accumulate different sets of allelic variants. They then migrate and mingle, and start interbreeding again quite randomly. The transfer of variant alleles from one population to the other is called *gene flow*.

So genetics is what rescued Darwinian natural selection from oblivion, forging the neo-Darwinian synthesis, which continues to provide a powerfully effective map for explaining the origins of all biological diversity on planet Earth, both now and historically.

Big ideas in science often benefit from scientists who are good at communicating the key results to a wider public, and the unusual J. B. S. Haldane did precisely that for the synthesis. Haldane has been described by Stephen Jay Gould as "independent, nasty, brilliant, funny, and totally one of a kind."[3] He learned Mendelian genetics as a boy by breeding guinea pigs and often served as one himself when he helped his father, who was professor of genetics at University College London. In one childhood episode, his father made him recite a long Shakespearean speech in the depths of a mine shaft to demonstrate the effects of rising gases. When the gasping boy finally fell to the floor, he found he could breathe the air there, a lesson that served him well later in the trenches of World War I.

Later Haldane himself quite often experimented using his own body, one time drinking a large quantity of hydrochloric acid to observe its effects on muscle action. I hasten to add that none of these experiments should be repeated by anyone reading this text, but it's perhaps not surprising that the writer Aldous Huxley incorporated Haldane into at least one of his novels as the archetypal eccentric scientist.

More relevant to our immediate topic is Haldane's ten highly mathematical papers published between 1924 and 1934, plus his influential book *The Causes of Evolution* (1932), in which he reestablished a central place for natural selection in the neo-Darwinian synthesis. It's interesting that Haldane comments in his 1932 book, "Criticism of Darwinism has been so thoroughgoing that a few biologists and many laymen regard it as more or less exploded."[4] This statement shows just how far the drift away from Darwinism had gone since 1882. But Haldane's aim was to resurrect Darwinism by showing that continuous, small-scale variation could also have a Mendelian basis, and especially that tiny selection pressures, working in a cumulative manner on such minor variations, could effectively explain evolution.

Haldane was a theoretical biologist who never did much fieldwork, but he did use the famous results of the biologist J. W. Tutt on the peppered moth. Increasing industrialization in Britain had led to a higher proportion of black moths, presumably because they were less visible to predators as they rested on sooty leaves. Haldane calculated that the observed increase of black moths from 1 percent in 1848 to 99 percent in 1898 required only a 50 percent higher survival rate of black moths over speckled ones. But if the increase was solely due to variation without selection, as the early Mendelians tended to argue, then this would require one in five moths to mutate from speckled to black, an obvious impossibility. Although more recent work has improved the quality of the data, Haldane's general point was a sound one.

Other influential biologists followed up in popularizing the new neo-Darwinian synthesis. Julian Huxley, brother of Aldous and grandson of Darwin's great defender, Thomas Henry Huxley, was the author of *Evolution: The Modern Synthesis* (1942), one of the most influential books on evolution in the twentieth century. He carried out famous studies on the Great Crested Grebe and on other birds that mate for life, developing ideas that Darwin himself had originally discussed on the evolution of sexual selection. Like Haldane, Julian Huxley was one of the biologists in the early twentieth century who restored a prominent role to natural selection in the evolutionary narrative.

Other key figures who helped establish the neo-Darwinian synthesis include the Russian, later to become American, Theodosius Dobzhansky, a committed Eastern Orthodox Christian who was a student of Morgan and first used genetics to investigate natural populations of *Drosophila* in the field. The title of one of his popular papers, "Nothing in Biology Makes Sense Except in the Light of Evolution," published in 1973, has become almost a mantra in the field of evolutionary biology. It's interesting to note how three of the great founders of contemporary neo-Darwinian theory—Haldane, Fisher, and Dobzhansky—represent such an interesting range in their own religious commitments. Haldane was the atheist albeit mystic Marxist; Dobzhansky, the Eastern Orthodox; Fisher, a committed Anglican who sometimes preached in his College Chapel in Cambridge—a good example of how scientists of any faith or none can contribute in the scientific enterprise to establish a common theory.

Today the neo-Darwinian synthesis continues to provide a powerfully effective map for explaining the origins of all biological diversity on planet Earth. It is neither a perfect nor a static theory, and the map of life that it provides will surely be adjusted and refined to incorporate new data as it emerges. Yet the evolutionary map, even in its present form, remains one of the most stunningly successful theories in the history of science.

GENETICS AND DARWINIAN EVOLUTION TODAY

This historical introduction should make clear that evolution is currently understood to be the result of two steps or processes, both of which are critical for evolution to occur.

First, there is the generation of genomic diversity by all the many different mechanisms we have reviewed so far. There is the important mechanism of recombination during meiosis, when segments of the paired chromosomes are exchanged during the formation of the sex cells. Sexual reproduction itself leads to massive amounts of variation in a population as the different chromosomes from two parents are mixed and matched to generate different progeny. Genetic novelty might also come from any or all of the mechanisms that geneticists have now been able to list in some detail. These include point mutations, insertions and deletions, retrotransposons (the jumping genes), gene duplication and divergence, genome duplication, chromosomal mutations, retroviral insertions, lateral gene transfer, and very occasionally, the import of a complete external genome from some other organism. And let's not forget gene flow, described above.

The second main step, natural selection, operates to test the variant phenotypes generated by the different genotypes in the workshop of life, so that a population of organisms becomes better suited over time to its environment. It works because the genes of individuals that produce more offspring are better represented in the next generation than the genes of individuals that produce fewer offspring. Thus, a genome that, on balance, aids an organism's survival and reproduction will be "selected for," being more fit, whereas a genome that, on balance, hinders survival and reproduction will be "selected against," being less fit.[5] Selection acts on the individual, which is a product of its many genes and their interactions with each other—some advantageous, some disadvantageous, and some neutral. The key criterion is that of reproductive success: How many progeny representing particular ensembles of genetic variants in their genomes are generated, and how many of these pass their genomes on to subsequent generations?

The term "survival of the fittest" has sometimes been used to describe natural selection, but is not very accurate because survival is not really the main point in this process. Of course, if an animal or plant does not survive then it won't reproduce, but the key point about natural selection is the successful reproduction that allows an individual's genes to be passed on to the next generation.

Evolution is thus a two-step process: generation of diversity followed by rigorous filtering of that genetic diversity in the workshop of life. The great majority of genetic changes, if not neutral, are likely to be maladaptive, so they disappear from the population after some generations—or even immediately if they are lethal. On the other hand, the few beneficial changes that readily pass through the filter of natural selection spread throughout an interbreeding population as they bestow reproductive benefits on their recipients. Alleles that increase fitness in this way are said to become "fixed" in the population once all individuals possess the same allele.

At the same time, the effects of each allelic variant of a gene are defined by the company it keeps. It takes many different genes to organize development, construct a complex metabolic pathway inside cells, or build a limb or an eye. A variant allele that might bestow benefits when its effects collaborate with ninety-nine other alleles to produce a certain outcome may no longer be beneficial when present along with a slightly different set of ninety-nine alleles.

The metaphor of the selfish gene has been used to present a gene's-eye view of evolution, in which the gene is envisaged as a selfish replicator occupying a "survival machine" that only exists for the gene's benefit to convey it onward to the next generation.[6] The metaphor has some use in that it draws attention to the importance of each variant allele in building organisms, but overall it is somewhat misleading. In reality each gene is dependent on the actions of many other genes. Genomes provide systems for building organisms in a cooperative, interactive way; there is nothing "selfish" about that. And the functioning of one gene may change considerably depending on the presence or absence of variants of other genes. So the metaphor of genes as "cooperators" might be more accurate as a way of describing how they collaborate together in real life to carry out complex, interacting functions.

We can picture the way in which genes cooperate like an orchestra. In a real orchestra, a musician playing a particular instrument may flourish if the environment provided by the rest of the orchestra generates the best interactions to play a great symphony, but equally the sound of even a very gifted player is quickly spoiled if the rest of the performers are off-key. So it is with genes: the very same gene can exert a rather different effect in building and running the organism depending on the company that it keeps. Strictly speaking, therefore, an allele cannot be defined as "beneficial" or "deleterious" as if this were its permanent characteristic. As already emphasized, it all depends.

Genomic data have also revealed that the notion of a "neutral" allele—that is, a variant gene that supposedly neither increases nor decreases fitness—is also problematic. You might have thought that mutations that cause no change in the amino acid sequence of proteins would be "neutral" rather than having any influence on selection. After all, exactly the same protein is produced. However, such mutations can in fact make a difference to evolution.[7] Scientists have discovered, in fact, that these so-called neutral alleles can assist in making the translation of genetic information into proteins more or less efficient.

The impact of variant alleles may also vary greatly depending on whether they are present in either one or two copies: this is often the reason why a genetic disease will appear in some human carriers, but not others. Sometimes the heterozygous condition, in which a mutant allele exists on one of the pair of chromosomes but not the other, is actually beneficial for its carrier, even though the homozygous condition for the same mutant allele is profoundly deleterious. This kind of situation leads to a balance of genetic variants in the population in which a certain proportion of heterozygotes is maintained, due to the advantages bestowed, despite the fact that occasionally heterozygotes will mate, with a one-in-four chance of any one of their progeny having the homozygous deleterious condition.

An actual example of this scenario is provided by an allele that helps protect against malaria. Each year, about 400 million people contract malaria, and 2 million to 3 million die, the majority of whom are children. Malaria is caused by a parasite that infects red blood cells and feeds on hemoglobin, the protein responsible for transporting oxygen in our blood. In parts of the world where malaria is common, such as some African countries, a particular allele of the hemoglobin gene is also common. This "hemoglobin S allele" contains a single-point mutation that makes the hemoglobin protein prone to clumping, which in turn makes the normally disc-shaped red blood cells become elongated and rigid, or sickle-shaped, as figure 14 illustrates. The parasite finds it hard to feed and reproduce in the sickle cells, so the risk of malaria is reduced. A double dose of hemoglobin S ("homozygous") means that the sickled blood cells clump even more, blocking capillaries and preventing oxygen from being carried to the body's tissues. Imagine pushing a bunch of oranges down a pipe with lots of twists and turns, and then try it with bananas—it gets much harder. This homozygous condition is called sickle-cell anemia, and about 80 percent of people who have it die before reproducing. But natural selection doesn't weed the hemoglobin S allele from the population, even though it's very deleterious in the homozygous condition, because it bestows the partial protection against malaria to heterozygotes.

In practice, natural selection does not weed out all deleterious alleles, even when they bestow no particular advantage upon the organism in the heterozygous condition, as in the example above. The reason is that deleterious alleles often hitchhike along with a beneficial allele located nearby on the chromosome. We say that alleles are "linked" when they tend to travel together during recombination and get inherited as a "package." Of course, if the net effect of the deleterious alleles outweighs that of the beneficial allele, then the overall fitness of the organism will decrease, and this genotype will tend to be weeded out by natural selection. But if the deleterious allele is relatively mild in its effects, then hitchhiking will be successful.

Whether variant alleles are beneficial depends not only on the genetic company they keep at the genomic level, but also on the environment of the organism that they help to build. Think about the anteater, which needs a long snout to delve down into big anthills and narrow crevices to fish out those delicious little ants, finding his snout getting longer and longer over succeeding generations only so long as its advantages outweigh its disadvantages. Crevices

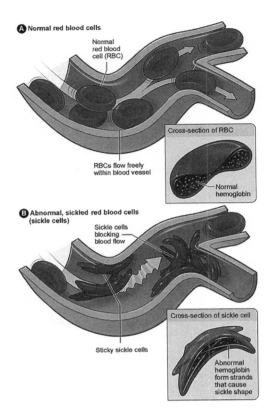

FIGURE 14. Normal (A) and sickled (B) red blood cells. Sickle cells are a result of a mutant form of hemoglobin that is less flexible than normal hemoglobin and distorts the shape of the blood cell. One copy of the mutant form of hemoglobin provides some protection against the parasite that causes malaria, but two copies of the mutant hemoglobin gene cause sickle cell disease in which blood circulation is impaired by the profusion of sickle-shaped cells. Reprinted by permission from the National Heart, Lung, and Blood Institute, http://www.nhlbi.nih.gov/health/dci/images/sickle_cell_01.jpg.

are only so deep, and at a certain point the snout gets so long that it slows down escape from predators. So a kind of equilibrium is reached in which genomes build snouts of just the right length to do the job that needs to be done without being a handicap: anteaters with that optimally useful length of snout will eat lots of ants, flourish, and have loads of offspring, passing on their useful genomes to succeeding generations. Notice that this natural selection process is very different from the Lamarckian idea that the anteater tries hard to get ants just out of reach, thereby lengthening her snout, so passing her long snout on to her offspring. That is not how it happens!

But if the supply of ants suddenly dries up for some reason, an anteater with a long snout may be in trouble. A different selection pressure now begins to operate, and adaptations that enable anteaters to successfully use alternative food sources might develop. The ecological niche defines what kind of adaptations will develop. An anteater's long snout is not going to be of much help to polar bears.

Natural selection is generally a rather slow process: steadily winnowing, shaping, weeding, and revising genomes along with their attendant phenotypes. Biologists in a hurry to find answers turn to organisms that reproduce fast, such as bacteria that can divide every twenty minutes if fed well, and examine their evolution in the laboratory.

Rich Lenski and colleagues, then at the University of California, did exactly this on February 24, 1988. They started growing a series of twelve populations of the bacterium *Escherichia coli*, all derived from a single bacterium and fed using glucose. The evolution of different strains of these bacteria from the original parental cells has now been tracked for a period of more than twenty years.[8] Each day about half a billion new bacteria grow in each flask, involving the replication of the same number of bacterial genomes, and in total about a million mutations occur in each flask as the bacteria divide. Since there are only about 5 million base-pairs in the bacterial genome, this means that every few days virtually the whole genome will be subject to genetic analysis to see whether any of the new mutations might be useful. In practice the vast majority are not, but new mutations occasionally come along that provide some growth advantages.

Every night the bacteria run out of their glucose food source and become dormant, so bacteria that cope best with this changing environment have a big advantage. The next day about 1 percent of the culture from each flask is used to start a new culture with a new supply of glucose. Most of the beneficial mutations that occur provide up to a 10 percent growth advantage, and such mutations spread rapidly through the population as the progeny carrying the mutation have this modest growth advantage. What Lenski found was that the evolution of the different flasks of bacteria, as measured by their growth, developed not in a smooth trajectory but in a series of abrupt jumps as advantageous mutations took over the population. These are not quite the "saltations" that were popular in the early twentieth century, but certainly highlight the fact that even a small change in one or a few genes can make a big difference to the success of the organism.

After more than a decade of subculturing the twelve flasks, something rather extraordinary happened at generation 33,127. One of the cultures "discovered" how to use citrate as a food source, a chemical used to stabilize the pH and so present in all the flasks since the beginning. It was like a population of cats suddenly taking a liking to whisky (assuming an unlimited supply), and it gave this population a huge growth advantage as it was no longer dependent upon glucose as a food source. This critical event happened in only one of the twelve flasks and it took more than ten years to show up. Further analysis revealed that the capacity to use citrate could not evolve all in one step, but took three different mutations to achieve. The two "background" mutations had to occur first, and the third critical mutation then enabled the complete ensemble of three mutations to allow the use of citrate, thereby opening up a whole new way of living for the colony. In fact, what has happened is that 99 percent of the colony uses citrate, whereas about 1 percent have become "glucose specialists," stubbornly refusing to forsake their original food source.

This wonderful experiment highlights the need for multiple cooperative alleles to emerge together in order to generate a selective advantage, and also how one key mutation can open up a whole new landscape of evolutionary possibilities. Indeed there are many examples in evolution in which beneficial alleles can rapidly sweep through a population, producing sudden bursts of evolution, sometimes associated with speciation.[9]

The Red Queen

The Red Queen Hypothesis has proved to be a fruitful idea when considering the tempo of evolution. The term derives from Lewis Carroll's *Through the Looking-Glass*, in which the Red Queen remarked, "It takes all the running you can do, to keep in the same place." Applied to evolution, the idea is that organisms in active competition with other organisms have to keep evolving just to keep pace with their environment. This is strikingly apparent in the "arms race" of host-parasite relationships in which the host is evolving to protect itself against the parasite and the parasite is seeking to outwit the host.

Once again, bacteria provide us with a vivid example of how this happens in practice. Parasitic viruses that infect bacteria are called Phage, and scientists have studied Phage Φ2 (pronounced "Phage Phi 2")[10] and the bacteria that they infect in order to better understand their evolutionary arms race. The rate of evolution as measured by genetic variation was far higher in the Phage when it was cocultured with the bacteria continually ("coevolution"), rather than when different batches of the same genetically identical bacteria were independently infected with the Phage for the same time period. Furthermore, four Phage genes in particular evolved much faster than other genes (by mutating), showing that they were actively being selected. Perhaps not surprisingly, these were all genes involved in host infection, for the Phage can only replicate by first infecting the bacteria. So coevolution seems to stir things up, preventing organisms from getting stuck in a rut and driving evolutionary change.

But the bacteria still stayed bacteria in this experiment, just as they did in Lenski's culture flasks over decades of evolution. To see how genetics is involved in the transition of one species into another, we have to stand back a little and look at the evolutionary tree of life as a whole.

Notes

1. Francisco J. Ayala, *Darwin's Gift to Science and Religion* (Washington, DC: Joseph Henry Press, 2007); Jerry A. Coyne, *Why Evolution Is True* (New York: Oxford University Press, 2009); Richard Dawkins, *The Greatest Show on Earth: The Evidence for Evolution* (New York: Bantam Press, 2009).
2. A more detailed explanation of the various nuances of the term "fitness" in evolutionary biology may be found in H. A. Orr, "Fitness and Its Role in Evolutionary Genetics," *Nature Reviews Genetics* 10 (2009): 531–39.
3. Stephen Jay Gould archive, http://www.stephenjaygould.org/people/john_haldane.html, accessed November 19, 2010.
4. John B. S. Haldane, *The Causes of Evolution* (London: Longmans, Green, 1932), 32.
5. See note 2, this chapter.
6. Richard Dawkins, *The Selfish Gene* (Oxford: Oxford University Press, 1976).
7. L. D. Hurst, "Genetics and the Understanding of Selection," *Nature Reviews Genetics* 10 (2009): 83–93.
8. J. E. Barrick et al., "Genome Evolution and Adaptation in a Long-Term Experiment with *Escherichia Coli*," *Nature* 461 (2009): 1243–49; T. Chouard, "Revenge of the Hopeful Monster," *Nature* 463 (2010): 864–67.
9. M. Pagel et al., "Large Punctuational Contribution of Speciation to Evolutionary Divergence at the Molecular Level," *Science* 314 (2006): 119–21.
10. S. Paterson et al., "Antagonistic Coevolution Accelerates Molecular Evolution," *Nature* 464 (2010): 275–78.

Justin L. Barrett
Cognitive Science and Religion

AS BARRETT TELLS US in his book *Cognitive Science, Religion, and Theology*, this emerging research field is not mere brain science. Instead, it is "the science of the mind and how we think." The mind and thinking take in a wide territory: perception, attention, memory, conceptualization, communication, reasoning, learning, decision making, and imagination. These human qualities develop as we move from infancy to adulthood. In Barrett's cognitive model (which differs from Brown and Jeeves' "emergence" model), the biological makeup of humans gives them "natural" beliefs from the start. These are refined by "reflection," which over the ages has produced cultures and theological systems to explain the world in detail. In science, the new field of the cognitive science of religion looks closely at these processes in the laboratory and in the field, conducting research that overlaps with anthropology, philosophy, and even computer science, which now explores cognitive functions of mind. "Cognitive science can be a fruitful dialogue partner with religion," Barrett says. Science finds that human beings hold a "rudimentary sense of the divine," which seems rooted in basic human psychology. Historically, this sense is elaborated as religion, part of the "cultural scaffolding" that makes up societies. Although cognitive science is hard science, and skeptical until evidence is shown, its interest in human belief allows that even "theologians can help guide cognitive science into important new problems."

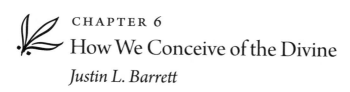

CHAPTER 6

How We Conceive of the Divine

Justin L. Barrett

WHEN VISITING INDIA, one cannot miss the colorful diversity of shrines and temples devoted to the various gods. You will find images of multi-armed human forms—elephant-headed Ganesha, the monkey-god Hanuman, the god Vishnu in the form of a fish or a boar, and hundreds of others. These various images correspond to as many (or more) different conceptions of the divine—and this is only in Hindu India and only considering those beings that are objects of worship and not countless demons, spirits, fairies, and other superhuman beings. If we consider all of the superhuman concepts from the world over, how can any general patterns be drawn?

In this chapter, I use a "cognitive science of religion" (CSR) approach to understanding why such concepts of gods are common around the world. Current research is suggesting that human beings, because of the way their minds ordinarily develop, find it natural to believe in divine activity, and they do so "nonreflectively," that is, almost automatically before there is time to reflect. Even after reflection, many of these beliefs continue. As I show, some god concepts have turned out to be more conceptually effective for injecting meaning into life's events. By the end of this essay, we'll have enough groundwork to consider whether cognitive science explanations for why people tend to believe in gods undercut justification for these beliefs.

WHAT IS A GOD?

One of the difficulties in studying how people think of gods is even identifying what counts as *god, divine, superhuman,* or *supernatural.* All of these terms are contested, and their use reflects a particular perspective or theoretical framework. Rather than try to define "gods" from a theological perspective—a perspective that is bound to be disagreeable to someone from a different theological vantage point—here I will treat *gods* and related terms from a cognitive scientific perspective. "Gods," here, refers to counterintuitive intentional agents

- ▶ That a group of people reflectively believes exists
- ▶ That have a type of existence or action (past, present, or future) that can, in principle, be detected by people
- ▶ Whose existence motivates some difference in human behavior as a consequence

Something is counterintuitive when it has a property that violates our everyday sense (our intuition) of how things normally exist. Figure 15 shows the normal intuitions that human beings have about the "ontology," or ways of existence, that we see around us. So, for example, a living cow with metal internal parts is "counterintuitive" because the expectation of having organic innards is intuitively applicable to Living Things (see figure 15), and this expectation is breached here. A tree that listens to people's conversations is counterintuitive because the expectation set Mentality (see figure 15) has been transferred to an ontological category to which it does not belong. *Counterintuitive,* then, as a technical term motivated by current understanding of natural human conceptual systems, maps roughly onto how people often use the terms *supernatural* or *superhuman* without running aground on the problem of specifying what is natural and what is "super" or above humans. (What if the divine is part of nature? What if a god is decidedly subhuman in some respect?) Such a definition of a god, as it appeals to cross-culturally recurrent features of human cognition, avoids the problem of being nonapplicable to other times or places.

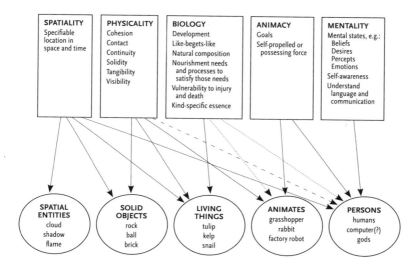

FIGURE 15. How activation of expectation sets comprise intuitive ontologies. Arrows indicate the relationship between expectation sets and intuitive ontologies. Dotted lines represent that Biology need not but may be active for Animates and Persons. Dashed lines represent that, provided intuitive dualism is correct, Persons need not include an activation Physicality.

Gods, as defined in this way, are certainly not the only concepts that have occupied theologians or thinkers concerned with the ultimate concerns, the meaning of life, or spiritual pursuits. If one wants to see Paul Tillich's idea of God as an abstract "ground of our being"[1] or variants on the theme that God is some kind of impersonal force much more like that portrayed in the *Star Wars* movies than that in the Bible, then the cognitive approach I am advocating will disappoint. From this cognitive perspective, such nonagentive ways of viewing "God" or "the divine" do not fall into the same category with Allah, Shiva, or Zeus as conceived by ordinary devotees.

Perhaps surprisingly or disconcertingly, demons, ghosts, ancestors, and other traditional religious beings are gods under this framework. Whether a god is good or bad, powerful or feeble, all knowing or laughably fallible, or worthy of devotion or scorn does not factor into this definition of gods. I regard this inclusiveness as a strength. Ancestor-spirits and nature-spirits just might be the most common focus of rituals and other "religious" activity in traditional small-scale societies, and their echoes are present in even the most "complex," urbanized societies in the form of ghosts and saints. An account of belief in gods that marginalizes these as "superstition" is unduly narrow.

A key question before the cognitive scientist of religion is to account for why gods are so recurrent across cultures. Whenever a particular type of idea or practice is widely recurrent across cultures, the human scientist should want to know why this is so. Simply answering "Because gods exist" is an insufficient answer. Quarks might exist but that does not mean quark-concepts are widespread across time and space. It might be that fairies do not exist and yet beliefs in such beings have stretched from China to Ireland. Whether something exists is a different question than why people believe that something exists. Cognitive scientists study such seemingly banal questions as "Why do people believe there is a three-dimensional world around them?" without resorting to "Because there is." Even if the actual existence of a three-dimensional world is patently obvious to us, there is still a cognitive problem regarding just how we come to find belief in such a world patently obvious.

A second important question is why do some of these gods become connected with belief systems or theologies concerning meaning and ultimate concerns? Fairies might meet the definition of a god and they might be widely recurrent, but they do not figure centrally in theological treatments of why humans are here, how they should understand themselves, and how they should live their lives. Why not? Why are some gods better candidates for providing meaning and becoming focal in religions than others?

WHY GODS?

Gods are common because of the operation of ordinary natural cognitive systems we use to make sense of the world and especially minded agents. The concert of various content-specific cognitive systems facilitates our belief in gods. We humans are eager to make sense of the world around us and have a number of early developing natural ways to do so. If something bears the marks of a bounded, physical object, then the Physicality and Spatiality expectation sets kick in, providing inferences, explanations, and predictions for the object's properties and movement. What if something does not appear to move in a way easily explained in terms of ordinary mechanics? As research has shown, it is our human nature to seek out intentional agents—particularly other humans—in the environment. Persons such as humans violate Physicality in some important ways, and this partitioning of things that wholly conform to Physicality from those that do not is a conceptual act in which even infants engage. Persons play by different rules than ordinary objects, and babies know this. Persons and their causal properties can account for events and states of affairs that the rules of Physicality cannot. A series of studies by the psychologist Deborah Kelemen has found, for example, that children naturally use design reasoning. Children often regard *someone* as a more sensible cause of natural states of affairs (such as features of mountains and camels) than some mechanical process.[2]

We have other natural causal strategies available, including essentialist and vitalistic reasoning that come from the Biology expectation set. Nevertheless, given persons' (apparent) potential ability to explain a broad range of events and states that are not obviously the result of these other forms of intuitive causal reasoning, and given the importance of person-based causation in human social life, we have a strong natural propensity to resort to person-based causal explanations when other forms of intuitive causal reasoning fail.

This natural tendency to resort to person-based conceptualization of events, states, and things may be an instance of *error management*. Humans and other animals have limited and fallible decision-making abilities that will at least occasionally lead to making errors, as in failing to detect food at hand or regarding something as edible when it isn't; or being scared of a harmless object or failing to notice a dangerous predator nearby. Because errors happen, in many cases it would be advantageous for an organism to err in one direction instead of the other depending upon which kind of mistake is more costly for the organism. For instance, the cost of mistaking a poisonous mushroom for an edible mushroom could far outweigh the benefit of finding lots of edible mushrooms. This observation applied to human cognitive systems entails that humans will have some tendencies that are tuned in a particular "safer" direction.[3] Anthropologist Stewart Guthrie has argued that the human perceptual and conceptual tendency to see human-like agents and agency everywhere—even in situations we later recant—is one such better-safe-than-sorry tuning.[4] I have labeled this cognitive system *HADD* for *hypersensitive agency detection device*.[5] Evidence that humans have such a device is strong.[6]

Cognitive scientists have demonstrated repeatedly that, from infancy, movement that looks self-propelled and goal-directed activates thinking about objects as agents, and further can trigger attribution of mental states, beliefs, desires, and sometimes even personality and social roles. Famously, psychologists Fritz Heider and Marianne Simmel showed American female college students a film of geometric shapes moving in and out and around a broken rectangle and then asked them to recall what they saw. Instead of producing simple factual descriptions such as, "The large triangle moved toward the circle and stopped just to its right. The large triangle then moved out of the rectangle and toward the smaller triangle and circle . . . ," observers mostly described the movement using personification—for instance, describing the smaller triangle and the circle as friends and the larger triangle as their enemy, and them wanting to escape, and the large triangle becoming angry.[7] This and subsequent studies illustrate that our perceptual and conceptual systems readily attribute minded agency with little provocation—even in conditions in which we reflectively think that such attribution is in error.[8] That is, we have an agency detection system that produces false positives (or at least we reflectively believe it to do so). Guthrie has argued that this tendency to find humanlike agency even where it does not actually exist is a primary generator of belief in gods. We have *HADD experiences*, experiences in which we detect agency for which the type of agency is unclear, and then sometimes regard them as evidence of a god or gods. Our natural cognitive systems find minded agency even where there is not any, and are likewise attracted to intentional explanations for natural events and states of affairs. Gods, by these lights, are false positives.[9]

Even if one is uncomfortable with Guthrie's decidedly dismissive approach concerning the possible existence of one or many gods, his general point is helpful. Our natural cognition readily applies purposive and mentalistic construals to a broad range of objects, states, and events even given only ambiguous evidence that mental agency is in fact at play. If one has been exposed to a god concept—and recall that in principle gods' activities are detectable in the world—one's agency detection system has a good likelihood of detecting evidence of the god's activity. Similarly, given this agency detection system that is more forgiving of false positives than failures of detection, one will occasionally encounter events or conditions that *seem* to cry out for an explanation in terms of the activity of an intentional agent and a regular human or animal will clearly be insufficient. Such events might occasionally lead to the postulation of a god (or support the existence of a known god).

To illustrate, recently when I arrived at home and prepared to open my back garden gate, the door slowly swung open just in front of me, without me contacting it—not a normal occurrence! While it opened, I assumed that some*one* on the other side heard me coming and opened it for me, but when it was fully open, I saw no one who could have opened it. I checked behind the gate and saw no one. Strange. My agency detection system registered agency (I had a HADD experience). I assumed a human agent was present, but then no human agent was present. Who did it? We probably all have such experiences. Most of the time we shrug it off, but perhaps occasionally such experiences reinforce beliefs in ghosts, spirits, and the like.

Similarly, as Kelemen's work suggests, we seem to have a conceptual bias to see natural things in terms of their function or purpose, and have a tendency to connect this apparent purpose to the activity of someone—an intentional agent or person. Finding purpose in the natural world and eagerly attaching purpose to intentional agency would provide impetus for believing in beings that can account for the apparent purpose or design in the natural world. As humans and animals are rarely good candidates, conceptual space is open for gods to fill this role.

Taking Guthrie's and Kelemen's work together, we see that humans have natural, intuitive impetus for postulating gods. Events and things in the world appear purposeful, designed, or otherwise the product of minded, intentional activity. It has been speculated that events of unusual fortune or misfortune likewise provide motivation to consider the existence and activities of gods. When some improbable event happens that seems meaningful, we might readily assume someone, perhaps a divine someone, is responsible.[10]

Such thinking could be reinforced by moral intuitions as well. Psychologists have discovered some evidence that we easily think about the world as operating on some kind of reciprocity principle, what is called *just world reasoning*.[11] If someone does something morally wrong, something bad is more likely to befall him. But why? One way of theologically elaborating this intuition is to postulate a punishing and rewarding force such as karma, as we see in many Asian religions. Alternatively, a fairly intuitive account would be that someone knows about the wrongdoing and punishes it. As gods might know even what is done in secret, and can use natural forces and things to reward or punish, gods may serve an explanatory role in these cases of unusual fortune or misfortune. Even gods that have little concern about human interactions may be vengeful when humans trespass against them.

The natural tendency to see agency around us, to see purpose in the world, to demand explanation for uncommon fortune or misfortune, and to connect fortune or misfortune to reward and punishment may conspire to make gods readily understandable and provide impetus for entertaining their existence and activities. One further set of considerations deserves mention as well.

How Dead People Become Gods

Many if not most of the world's gods (broadly construed) bear some relation to deceased humans. Ancestor-spirits and ghosts were, at one time, humans. In small-scale and traditional societies, warding off ghosts and malevolent spirits, propitiating the ancestors, or garnering the support of (deceased) saints often takes on far greater importance in regular practice than concerns about creators or cosmic deities. What might account for this cross-cultural recurrence?

At this point the answer is largely speculative. If we accept that our natural cognitive equipment makes us intuitive dualists (as suggested by Yale psychologist Paul Bloom's research), then the idea that an immaterial something—mind, body, spirit, or some combination thereof—is left behind after death is not radically counterintuitive. Even if such dualism is not an intuitive default but is merely an easily accommodated idea, the point remains that mind- and soul-related ideas are not difficult to decouple from bodily reasoning, and the fact that someone's body has stopped working need not conceptually turn off reasoning about the person's thoughts, feelings, desires, and other properties. Further, as Pascal Boyer has noted, with someone intimate, we know a lot about their tastes, desires, preferences, personalities, and the like;

and upon death these mind-based properties remain untouched. Our theory of mind system, informed by such information, continues generating inferences and predictions even after someone has died.[12] Add to these considerations the occasional, emotionally gripping experiences related to the deceased, such as dreams, hallucinations, or strange sounds, which trip our enthusiastic agency detection system (HADD), and you have a recipe for supposing that the recently deceased is still active without his or her body. Indeed, the distinct sense that someone passed away is still somehow present is not uncommon even among people who explicitly reject belief in ghosts and spirits of the dead.[13] As the dead no longer have visible, physical bodies, they may satisfy the search for an intentional agent in the many cases in which our agency detection system detects agency but it is clear a visible human or animal could not have been responsible (HADD experiences). It follows that little cultural encouragement is needed to develop the idea that some form of the dead is still around and active among the living in some cases.[14]

MINIMALLY COUNTERINTUITIVE IDEAS AND CULTURAL TRANSMISSION

Thus far I have been stressing the content-specific natural cognition that likely encourages thinking about gods and entertaining their existence and activity. If these dynamics characterize many people's automatic thinking in ordinary human contexts, we would expect that ideas about gods would be readily generated by individuals (if only in a piecemeal fashion) and transmitted from person to person, perhaps gaining elaboration. As god concepts have a strong intuitive foundation, conforming well to nonreflective beliefs, they will be strong candidates for reflective beliefs as well. Research in cognitive science has shown that if something is easier to bring to mind, even if only because of repeated exposure—that it has strong accessibility—then it usually breeds believability.[15]

Though I have been stressing intuitive impetus, a god concept need not be wholly intuitive but might deviate a bit from the expectation sets (see figure 15). As long as a concept is not *too* counterintuitive, the concept need not suffer in terms of how easy it is to conceptualize, remember, and communicate. Consider a spirit that can read minds. Mind reading might be counterintuitive, but a disembodied mind having the additional feature of being able to know what people are thinking is not particularly cumbersome. It is only modestly counterintuitive and so is easy to think and talk about. Some experimental research shows that concepts that are only slightly counterintuitive are as or more memorable and transmittable than wholly intuitive ones.[16] If so, then their good potential for being communicated will increase with how often they are talked and thought about. Thus, they will be more frequently encountered and, in turn, become more familiar and accessible, leading to believability. Being mostly intuitive and just slightly counterintuitive poses no special problem for a religious concept and may actually be an asset.

Not all god concepts are only slightly counterintuitive, however. Many have features that deviate radically from natural expectations and are highly counterintuitive. Consider the idea of a god that knows all, has no location whatsoever, exists in a different time (or no time at all), and yet, can interact with the world. This approximates a view of the divine that we see in some theological treatments in Judaism, Islam, Hinduism, and Christianity. What then? Such concepts are likely to require special cultural scaffolding to aid their transmission: special artifacts, institutions, practices, or other devices that help people learn and use these more complex concepts.[17]

GODS AND MEANING-MAKING

Some Westerners think of gods in terms of their ability to give existential meaning to life, to account for why we are here, what makes life worth living, and how we should conceive of ourselves in relation to other humans, the rest of nature, and the cosmos. Though some gods might help with the big existential questions, perhaps most are more modest in their meaning-making. *Why are my chickens laying poorly? Why was this fishing trip so successful? How can I get the spouse that I want?*

As noted previously, intentional beings (Persons) can help explain and predict a broad range of phenomena either as proximate or more distant causes. As concepts, they possess what Pascal Boyer has called *inferential potential*, the ability to generate a broad range of ideas, inferences, explanations, and predictions about issues that matter to people. That is, a concept with inferential potential has the potential to enable people to *infer* or draw conclusions across a broad range of concerns.[18] Appealing to the activity of an average person (let alone a super person) can account for a greater variety of things than appealing to the average rock, shrub, or beetle. Not all person concepts—human or gods—have the same inferential potential, however. We naturally learn about human causal limitations and thus understand, for instance, that positing human activity cannot make meaning from a freak thunderstorm that destroys my crops. Humans have a restricted range of things they can do and, thus, can explain.

Boyer has argued that gods—particularly the ones that attract a lot of attention and behavioral investment—are those whose counterintuitive property or properties give them greater and broader inferential potential in domains of human concern. For instance, by virtue of being invisible, being able to read minds, or seeing all, a god can have access to strategic information about who plans to do what to whom, and may act on that information or share it. Different counterintuitive properties, such as failing to exist on Wednesdays or only knowing languages that cannot be used, yields little gain in inferential potential. Such beings will be unable to importantly bear upon human concerns or experiences. They are not terribly interesting or worth talking about, and so will be less accessible, and hence, less believable. Consequently, such beings will usually fail to persist as shared ideas and will rarely (if ever) become recognized as gods.[19]

Even successful gods vary in the ability to make meaning. Some local forest spirits might be very powerful and inferentially important when it comes to reasoning about what goes on in a particular part of the forest, but completely irrelevant when drawing inferences about domestic affairs, whereas the opposite would be true of domestic spirits. Gods vary in the meaning they can help make. Those with broadest ranges of activity—both spatially and in terms of kinds of things they can and cannot do—will generally have the greatest ability to serve as meaning-makers. A cosmic creator who can act in essentially any domain of life has more meaning-making reach than the ghost haunting the house on the corner and, therefore, is more likely to come to mind, be talked about, and be accessible and believable (all else being equal).

I raise these observations concerning how gods are recruited to make meaning because, from a Western perspective, it is commonly assumed that gods are all about meaning and that all gods are equal in this regard. Westerners sometimes talk about religions as fundamentally existential meaning-making systems. The suggestion is that those "spirits" or "ghosts" that do not serve this function are not really a part of religion, but are merely part of local superstition and should not be called *gods*. In terms of conceptual properties, however, a clear dichotomy does not exist.

Reasons to Doubt the Existence of Gods?

Gods come in dazzling diversity, from ancestor spirits to cosmic creators, but amid this diversity we can see that a small number of conceptual factors make an idea more likely to become entertained and believed in as a god. A strong god candidate is a modestly counterintuitive intentional agent or person, because such persons have great inferential potential while not being too complex or difficult to understand and communicate. Successful gods also tend to produce actions that are detectable in the world, either through our agency detection system, or because they account for apparent design and purpose in the natural world. Beings that may be invoked as morally interested, and perhaps accounting for fortune and misfortune in terms of reward and punishment, may be especially successful. In short, successful god concepts need to be able to make meaning of life's events in relatively intuitive, straightforward ways. Observing these features of gods are not new. The novel contribution of cognitive science is that we can now better explain why these features are important for gods and why they tend to congeal.

How we conceive of the divine is shaped by the natural cognitive equipment we have, but by no means does this entail that beliefs in gods are mistaken or suspect on this basis. Merely providing a scientific account of such beliefs should not be taken as casting doubt upon their truth.

Elsewhere I have argued that identifying that a belief has a natural cognitive basis does not bear upon whether it is true, but may justify someone in holding such a belief to be true until sufficient reasons to the contrary arise.[20] That is, when it comes to naturally derived beliefs, we may treat them as "innocent until proven guilty." I believe the same principles apply here to belief about gods. Insofar as beliefs in the existence of gods are fairly direct, natural outcomes of ordinary human cognition, they may be regarded as justified until reasons arise to reject them.[21]

For instance, our natural tendency to see design and purpose in the natural world, and to regard that design and purpose as the product of intentional agency, seems to be cognitively natural and, hence, innocent until proven guilty. Likewise, in many situations, our agency detection system tells us that someone has acted and we are justified in this belief until we have good reasons to reject it. Note, however, that identifying the particular someone responsible for the perceived design in the natural world or as the agent whose action has been detected requires inferences beyond the immediate deliverances of natural cognition. The more inferential the steps, the less obviously justified such beliefs are. That is, intuitive religious ideas have stronger grounds for justified belief (all else being equal) than theological reflections and elaborations upon these beliefs. This conclusion is ironic in that theologians often regard themselves as in the business of sharpening and refining religious instincts and intuitions, much in the same way that philosophers intellectually sharpen and refine other classes of intuitions. Perhaps theologians do refine religious thought and consequently produce better intellectual products than what we receive through simple natural cognition. Nevertheless, such theological concepts may bear a greater burden of proof than simpler claims such as "That tree was deliberately brought about in some way by someone" by virtue of being less natural.

In this vein, philosophers have begun exploring whether doubt is cast upon religious beliefs or whether they are encouraged by cognitive explanations of religion.[22] Here I only want to dispense with three common arguments against belief in gods that arise in the context of cognitive explanations.[23]

One objection is that if belief in gods is produced by natural cognition, then they are not based on relevant evidence, and hence, are dubious. This objection mistakenly assumes that mature, reflective belief in gods is not based on relevant evidence. Though natural cognition facilitates reaching certain conclusions, it is an error to infer that natural cognition being part of belief formation precludes the use of evidence in the case of believing in gods. The account presented above clearly points to the role of agency detection and inferring design and purpose from *observations* of the world around us. Surely this constitutes one sort of relevant (even if insufficient) evidence. It might be that people accurately "read" intentional agency from the text of the natural world.

Alternatively, one might argue that perhaps our natural cognition does prompt us to believe in gods using evidence, but is simply unreliable in this regard. Our cognitive equipment might be fine for finding human agency with bodies and the like, but it produces belief in beings that do not exist such as ghosts and fairies and so it cannot be trusted when applied to gods. The difficulty with such a line is that it assumes that we *know* that there are no disembodied or invisible agents, so when our cognitive systems detect one, they do so in error. Such an argument is what is known as "assuming facts not in evidence" in American legal terminology or "question begging" in philosophical jargon. You cannot use the (alleged) fact that invisible agents do not exist as support for the claim that the cognitive system in question is error-prone and so undermines belief in invisible agents. You have assumed what you are setting out to reject.

Similarly, consider a household scale upon which we weigh things. I might suspect that it gives bad readings for any number of reasons, but I cannot determine that it is in fact error-prone without independently determining the weight of an object and then showing that the scale does not give its true weight. Just dropping a bag of potatoes on the scale and saying, "See, the scale says 10 kilos. That's not right. This scale is no good," would only be convincing if we knew already that the sack of potatoes does not in fact weigh 10 kilos. If our cognitive systems "weigh" our experiences and conclude that there is at least one god out there, we cannot take this conclusion as evidence that the cognitive systems are mistaken unless we have independent reason to think there are no gods. Indeed, normally we would regard such a "weighing" as evidence that there is indeed at least one god.

A third objection is a rejoinder to the first two. One may concede that the cognitive faculties that encourage belief in gods use inputs from the world around, and that we must leave open the possibility that at least one god exists, but these faculties result in many divergent beliefs in various sorts of gods. This divergence suggests error in the belief-forming mechanism, at least as applied to gods, and gives us reason to hold suspect the beliefs it produces. This more formidable objection still ignores an important point: natural cognition converges on the existence of at least one and possibly many gods. The precise character of the gods in question might vary, but it would be a mistake to take this variation as reason to reject justified belief in some kind of god. Imagine three people are out for a walk in the woods and they each think they see an animal in a distant thicket. One says it is a deer. One claims it is a fox. The third believes it to be a bird. Clearly, the belief-forming mechanisms in question are imprecise in this context, but should we conclude that they are not justified in believing they saw an animal of some sort?

Color perception provides another helpful analogy. Our cognitive system naturally gives us impressions about what color(s) various things are, but this fact does not preclude that there are disagreements in particular cases. A tropical fish might be described variously as red, fuchsia, magenta, pink, violet, or purple (just to name a few) depending upon the individuals' personal history, eyesight, and expertise. One person might even see no color at all (on this occasion) because he is among the roughly 7 percent of American males who are color-blind.[24] We would surely be in error to suppose that the lone person who sees the tropical fish and believes it to be violet does not have a justified belief on the basis of having an error-prone cognitive device. Likewise, people might disagree on the details of the gods in question, and a minority might fail to detect any gods at all, but it does not follow that those who believe in the existence of gods are unjustified in such beliefs. Many of these beliefs are likely to be false upon further, reasoned considerations, but as with belief in free will, souls, and consciousness, the naturalness of belief in gods should be taken as grounds for giving such beliefs the benefit of the doubt, as innocent until proven guilty and not as guilty until proven innocent.

Notes

1. Paul Tillich, *The Shaking of the Foundations* (New York: Charles Scribner's Sons, 1940).
2. Deborah Kelemen, "Are Children 'Intuitive Theists'? Reasoning about Purpose and Design in Nature," *Psychological Science* 15 (2004): 295–301.
3. A recent review and discussion appears in Ryan T. McKay and Daniel C. Dennett, "The Evolution of Misbelief," *Behavioral and Brain Sciences* 32 (2009): 493–510.
4. Stewart E. Guthrie, *Faces in the Clouds: A New Theory of Religion* (New York: Oxford University Press, 1993).
5. Justin L. Barrett, "Exploring the Natural Foundations of Religion," *Trends in Cognitive Sciences* 4, no. 1 (2000): 31. See also Barrett, *Why Would Anyone Believe in God?* 32–34.
6. Brian J. Scholl and Patrice D. Tremoulet, "Perceptual Causality and Animacy," *Trends in Cognitive Sciences* 4 (2000): 299–308.
7. Fritz Heider and Marianne Simmel, "An Experimental Study of Apparent Behavior," *American Journal of Psychology* 57 (1944): 243–49.
8. For a brief review of experiments on the topic, see Scholl and Tremoulet, "Perceptual Causality and Animacy," 299–308.
9. Guthrie's argument raises an interesting epistemological problem. How do we know that these are "false positives"? The agency detection cognitive system (or HADD) would have to be checked against a different, reliable system for identifying when agency has been accurately detected. Do we have such a system, or do we just have the one agency detection system? Maybe when we decide that we did not in fact detect intentional agency, it is at this point that the mistake is sometimes made. See Guthrie, *Faces in the Clouds*. Though evidence for some HADD-like system is strong, that it contributes to belief in gods requires more empirical attention.

   Guthrie uses different language in presenting his argument. He favors the term *anthropomorphism* to capture the tendency to attribute events and states to intentional agents. I avoid this term for three reasons. First, it carries the connotation that all intentional agency is humanlike, and this is an open question. In many cases it is not human agency that is being postulated but something different. Second, anthropomorphism is often associated with importantly different approaches such as Freudian or Piagetian, and can carry the assumption that the agency in question is thought to have a humanlike bodily form. My third concern is that the term *anthropomorphism* dodges an important theological matter: to what extent is human agency a reflection of divine agency—theomorphism—rather than the other way around?
10. More evidence for such a claim is needed, but the idea has been entertained by several cognitive scientists of religion; see Barrett, *Why Would Anyone Believe in God?*; Jesse M. Bering and Dominic D. P. Johnson, "'O Lord . . . You Perceive My Thoughts from Afar': Recursiveness and the Evolution of Supernatural Agency," *Journal of Cognition and Culture* 5 (2005): 118–42; Pascal Boyer, *Religion Explained: The Evolutionary Origins of Religious Thought* (New York: Basic Books, 2001); D. Jason Slone, *Theological Incorrectness: Why Reli-*

gious People Believe What They Shouldn't (New York: Oxford University Press, 2004).

Note that from a psychological perspective at least, when an improbable event happens, such as winning the lottery, pointing out that given enough time or enough people playing the lottery someone is bound to win does not necessarily remove the desire for an explanation of why. Sure, someone had to win, but why me? Sure, people get struck by lightning, but why me and why now?

11. Adrian Furnham, "Belief in a Just World: Research Progress over the Past Decade," *Personality and Individual Differences* 34, no. 5 (2003): 795–817; Melvin J. Lerner, *The Belief in a Just World: A Fundamental Delusion* (New York: Plenum Press, 1980).

12. Boyer, *Religion Explained.*

13. Jesse M. Bering, "Intuitive Conceptions of Dead Agents' Minds: The Natural Foundations of Afterlife Beliefs as Phenomenological Boundary," *Journal of Cognition and Culture* 2 (2002): 263–308; Jesse M. Bering, "The Folk Psychology of Souls," *Behavioral and Brain Sciences* 29 (2006): 453–62.

14. In one case in which a close friend died, I can remember hearing sounds—actual sounds, not hallucinations—that I automatically assumed were from him approaching, and had to consciously remind myself that he was dead and the sounds must have a different cause. Similarly, I had chilling dreams of him returning and trying to participate in regular life. It is easy to see how such experiences could set one to wondering whether the deceased really are still active and trying to interact, especially if people have convergent experiences that they share with each other.

15. Daniel Kahneman, "A Perspective on Judgment and Choice: Mapping Bounded Rationality," *American Psychologist* 58, no. 9 (2003): 699.

16. Justin L. Barrett and Melanie Nyhof, "Spreading Non-natural Concepts: The Role of Intuitive Conceptual Structures in Memory and Transmission of Cultural Materials," *Journal of Cognition and Culture* 1, no. 1 (2001): 69–100; Pascal Boyer and Charles Ramble, "Cognitive Templates for Religious Concepts: Cross-Cultural Evidence for Recall of Counter-Intuitive Representations," *Cognitive Science* 25 (2001): 535–64. See also Justin Gregory and Justin L. Barrett, "Epistemology and Counterintuitiveness: Role and Relationship in Epidemiology of Cultural Representations," *Journal of Cognition and Culture* 9 (2009): 289–314.

17. The metaphor *scaffolding* suggests that extra structures are needed to build up some ideas or practices that are not particularly natural (close to the ground). Further, scaffolding is not permanent and can be removed. Analogously, the cultural conditions or devices that help build up relatively unnatural beliefs and practices can be removed. Previously common knowledge (such as how to start a fire from natural materials) can vanish when the cultural scaffolding is removed.

18. Boyer, *Religion Explained.*

19. An exception to the "usually" in this statement is in cases where there is strong cultural scaffolding, such as institutionalized schools of thought that attempt to keep inferentially poor ideas alive. For more complete treatments of the role of inferential potential on the spread of religious ideas, see Boyer, *Religion*

Explained; Pascal Boyer, "Religious Thought and Behavior as By-products of Brain Function," *Trends in Cognitive Sciences* 7 (2003): 119–24.

20. Justin L. Barrett, *Cognitive Science, Religion, and Theology* (West Conshohocken, PA: Templeton Press, 2011); Kelly J. Clark and Justin L. Barrett, "Reidian Epistemology and the Cognitive Science of Religion," *Journal of the American Academy of Religion* 79, no. 3 (2011): 639–75.

21. And reasons often do arise. Suppose someone believes his god is fully visible under normal conditions and lives on the top of Mount Olympus. Repeatedly, he climbs Mount Olympus and fails to see the god. Either the belief that the god lives on Mount Olympus or that the god is fully visible or that the specified god exists becomes suspect.

22. David Leech and Aku Visala, "The Cognitive Science of Religion: Implications for Theism?" *Zygon* 46, no. 1 (2011): 47–64; Jeffrey P. Schloss and Michael J. Murray, eds., *The Believing Primate: Scientific, Philosophical, and Theological Reflections on the Origin of Religion* (Oxford: Oxford University Press, 2009); Kelly J. Clark and Dani Rabinowitz, "Knowledge and the Objection to Religious Belief from Cognitive Science," *European Journal of Philosophy of Religion* 3 (2011): 67–81.

23. For a brief, accessible treatment of these issues, see Michael J. Murray, "Four Arguments That the Cognitive Psychology of Religion Undermines the Justification of Religious Beliefs," in *The Evolution of Religion: Studies, Theories, and Critiques*, ed. Joseph Bulbulia et al. (Santa Margarita, CA: Collins Foundation Press, 2008), 365–70.

24. "Seeing, Hearing and Smelling the World: A Report from the Howard Hughes Medical Institute," Howard Hughes Medical Institute, http://www.hhmi.org/senses/b130.html, accessed November 24, 2010.

Javier Leach
Mathematics and Religion

$\forall x(x=x)$
$\forall x \forall y(x=y \to y=x)$
$\forall x \forall y \forall z((x=y \land y=z) \to x=z)$
$(1) \lnot \exists x\, GR(x,g)$
$(2) \lnot E(g) \to \exists x\, GR(x,g)$
$\therefore E(g)$

BESIDES OUR natural languages (English, Spanish, Hindi, for example), our modern world is shaped by three other types of language, says Leach in his book *Mathematics and Religion*. These are the languages of mathematics, the language of empirical scientific measurement, and finally metaphysical language, which is the symbolic language of religion. Each has a special purpose, but Leach tells us that in a global, pluralistic age, the formal language of mathematics is our most universal language. Mathematics arose more than five thousand years ago. Only in the last century has it advanced to the point of being "formalized," which has led to the dawn of the computer. The other result has been deeper discoveries of how logical systems work. Mathematics admits "different kinds of logic." Each one in turn is supported by a community. Despite this mathematical pluralism, and the realization that mathematics is an "open system" that often cannot be "decided," our bedrock for logic still is the human mind's consistency. When human life needs more than science, it has metaphysics, philosophy, and religion to comprehend a larger reality. Still, Leach says, these must be anchored in scientific knowledge. "I can separate mathematics from theology, but I cannot separate theology from mathematics," he concludes. "Mathematics and the empirical sciences are independent of religious belief, but theological reflection cannot do without mathematics and empirical science."

CHAPTER 7

On Math and Metaphysical Language
Javier Leach

IF WE WANT to keep our lives as simple as possible, mathematics and natural science offer a great advantage. Neither of them asks ultimate questions. Metaphysics, in contrast, is about ultimate things. For this reason, metaphysical questions may complicate our lives, but they also help us to resolve some of the deepest mysteries. Mathematics and natural science accept reality as a given fact; the questioning ends there. However, metaphysics asks why our minds are able, in fact, to understand the physical world and mathematics. Indeed, metaphysics asks why things exist at all. Why is there something rather than nothing?

Offering a more global and radical way of asking questions about reality, metaphysics points to the possible existence of an ultimate principle that justifies the existence of things in general. For most people, the metaphysical questions are unavoidable. But we also know from history that many kinds of perceptions of ultimate things exist, and therefore many kinds of answers to metaphysical questions. These questions and answers also form distinct communities, such as cultures or religions, just as we have seen in mathematics.

For example, one group may look at the metaphysical evidence in life and arrive at a basic intuition that the world exists on its own and for its own reasons. We may call this a *pantheistic* metaphysical view. Another kind of ultimate view may be called *agnostic*. It argues that we cannot know ultimate principles, so whether they exist is irrelevant. Third, a theist may believe that the universe exists because God exists and this Creator has made and maintains the universe. Finally, another metaphysical school may say that the ultimate principle in the universe is mathematics itself, as with schools such as the Pythagoreans and certain Platonists.

Metaphysics, of course, uses a different language from logic, mathematics, and natural science. This is the language of symbols that stand for ultimate realities or ultimate types of relationships. These symbolic words can range from *God* and the *cosmos* or *universe* to words found in mathematics, if that is deemed the highest reality. Whatever the language/symbol in metaphysics, it exceeds the meaning of the signs of mathematics and of the natural sciences. The word *number* is precise and definable in mathematics, but when used by the ancient Pythagoreans, for example, *number* not only referred to a mathematical object but also to the ultimate foundations of the world.

A scientific term is fairly objective when it speaks of a measurable object. A metaphysical term/symbol must be approached differently, however. A metaphysical or religious symbol is understandable only within a history, a tradition, and a community which uses that symbol. The symbol and its context provide its coherence. That context is empirical, for it is made up of history and tradition. What is more, metaphysical symbols can refer even to "Nature." But this is different from the physical measurements of natural science. Metaphysical symbols are, again, mostly determined by the community that uses them to speak of realities beyond what is empirical.

At one point in my life, my interest in both mathematics and theology was deepened by a particular type of metaphysical question. I had completed my advanced studies in mathematics and was pursuing theology when my professor stated the following idea: "The world is totally related to God, being totally different of him." On one hand, this proposition is about a kind of relation, which is what mathematics is all about. This relationship (regarding God) could be written in mathematical notations, just as we can convey the idea of 2 + 2 = 4 as a relationship of factors that, in the case of the world and God, do not add up, but one is totally related to the other, being totally different of it. But in a theological statement such as "The world is totally related to God, being totally different of him," more than just a formal (logical or mathematical) relationship is being conveyed.

This metaphysical, and therefore symbol-based, statement about God and the universe is possible because it is made in the context of a tradition and community—in this case, the Christian tradition. The symbols have had a resonance of personal meaning as well as a formal kind of logic. In metaphysical terms, the symbols speak of God's transcendence. They suggest that while the world depends on a Creator, the Creator does not depend on the world. For me, the idea had a transformative effect. Such symbolic meanings are about more than the mere objects related. The meanings can transform us and our communities.

Scientific Meaning and Metaphysical Meaning

In some ways, a scientific hypothesis also is a symbolic picture of how the world might be. But in science, we then try to test and verify that hypothesis. Under these tests, some hypotheses fail and others survive to be tested further. We test the truth of metaphysical and religious formulations differently. They are not subject to empirical testing as we would test scientific principles in a laboratory. Instead, metaphysical and religious formulations seem true to us when they offer an intuitive veracity and coherence in the context of the personal values that a group of people share. Those people outside the context naturally have a very hard time judging the intuitive truth and coherence of a metaphysical viewpoint. As years of interfaith dialogue have shown us, when a Christian states that Jesus is the truth and fullness of God, the Hindu, with another kind of tradition and community, cannot easily comprehend the statement.

Another difference between natural science and metaphysics is that in science, in principle at least, it is said that a hypothesis and theory are never absolute, but only the best approximation so far known in science. In metaphysical perception and language, however, the goal is to arrive at an absolute, a framework that can serve as a permanent vision of reality and the values that flow for it.

Because scientific hypotheses do not try to arrive at the finality known in religion, scientists can hold many metaphysical views. They can be agnostic, atheistic, or theistic in their fundamental beliefs. In other words, a scientist might easily understand the idea that God cannot be tested or proved because God, by definition, is not a mundane object. Religion offers a wider meaning to experience and history, even a timeless meaning.

When metaphysical symbols become articulated, they offer what we call *myths*, or grand narratives common to all religions and stories of the origins of the universe, or of how the human predicament came to be. (*Myth*, in this sense, does not mean an intentional falsehood, but rather a traditional story that tries to meaningfully unite the facts and mysteries of life.) A community shares these symbolic narrations and passes them on to subsequent generations. The heart of the myths, however, is a set of values that endure even when some of the religious language or narrative changes by its interaction with newly discovered facts about the world. For example, in Christianity, the stories of the human fall into original sin and the Resurrection of Jesus both convey narratives related to the human struggle with evil and hope for salvation, even though these myths can be told in different ways in Christian literature and theology based on our experiences in the world.

To be sure, science has its myths and its narratives as well. One of the most famous is the idea of progress, that science will always and everywhere find true answers and make the right decisions. At the heart of this scientific myth is the belief that everything changes, usually in an upward progression. But again, this symbolic kind of vision is hard to test in the laboratory. We seem to accept this myth as true, though, even though alternative views say that some things may not change, or that progress may be relative and elusive.

As increasing knowledge and communication in our world challenge many of our historical myths—religious and scientific—the myths that serve us best try to reconcile themselves to the empirical findings of science. Our most helpful myths cannot contradict the scientific data. Religion can accept the knowledge of nature that science provides. In turn, science hopefully can recognize its own necessary openness and its own myths and make room for a dialogue with the metaphysical languages that guide our human communities. This is more of a meeting of value systems than a battle over the facts of the world. The meeting is not a contest of pure logic, but a discussion of ethics, behavior, and history as well.

Nevertheless, there are fascinating times in our human experience when scientific and metaphysical concepts have overlapped in a very compelling way. As the rest of this chapter shows, two of those concepts are infinity and what has come to be called the ontological argument for the existence of an absolute—that is, for the existence of a supreme being.

The Infinite and the Ontological Argument

From the beginning of human knowledge, we have wondered how to conceive of the idea of infinity, something that goes on forever. We also wonder how this concept helps us in our lives, especially in mathematics and science. At the same time, we have pondered what it means for something to be the highest, greatest, or absolute in our world. As we will see, this concept was often expressed in the idea of God as "the greatest that exists" or "the greatest that can be thought." Remarkably, both of these ideas—infinity and the greatest realm of existence—have been parallel questions in math, science, and metaphysics. We begin with the concept of infinity.

Infinity: Mathematical, Empirical, and Metaphysical

For mathematicians, the infinite can be a sign with two distinct meanings: actual infinite and potential infinite. The potential infinite is a succession of mathematical objects that can be as big as we wish, because it does not end. For example, starting from a mathematical object that we designate as 0, we can construct its successor and designate it $s(0)$. Then we can define the successor to this object and designate it $s(s(0))$, and then to $s(s(s(0)))$, and so on. Thus, we can define the set $N = \{0, s(0), s(s(0)), \ldots\}$, which has as many elements as we wish. Then we can say that the number of elements of N is *potentially infinite*.

In contrast, we can speak of the *actual infinite* when we refer to all the numbers of N as a whole. When we consider N as a whole, we observe all the possible elements of N globally. (In the potential infinite, we consider N as a potentially infinite set observed not as a whole, but as a process of constructing its elements.) This distinction between potential and actual became important in the work of German mathematician Georg Cantor (d. 1918).

Cantor proved that the consideration of N as a "whole" (the actual infinite) leads us to consider other infinites that are "greater" than N, which usually are called *transfinites*. Some mathematicians could not accept this state of things. Now called *constructivists*, such mathematicians could admit only the existence of those mathematical objects that can be constructed based on finite symbols and functions. Hence, constructivists admit the existence of only the potential infinite in mathematical practice. They argue that the actual infinite cannot be constructed by finite means in a finite process.

Such disputes as this illustrate how mathematicians are allowed to use different systems, often simply as a matter of preference. If we choose classical mathematics (as Cantor had done), then we accept the actual infinite. A constructivist in mathematics accepts only the potential infinite. Both camps have a series of arguments to justify their choices in applying a mathematical approach to infinity.

In natural science, the concern is to arrive at an empirical idea of infinity that is useful for scientific hypotheses and research. Natural science draws on the mathematical idea(s) of the infinite, but tries to apply it in representational language that talks about the physical world. To begin, we must distinguish between the representational sign and the physical object represented. The infiniteness of time and space as a physical property of the world, for example, is often confused with mathematical infiniteness. To clarify, let's look at some of the physical descriptions of infinity.

Natural science has always struggled with the problem of trying to measure something that is infinitely small or infinitely large. Some physical cosmologies have seen time and space, for example, as infinitely large, and thus unmeasurable as an empirical object. However, current physics tends to think of the physical universe as being a limited object. In the big bang theory, for example, time originated at the original explosion, and time will end after a certain period of expansion of the physical universe. Indeed, once the expansion is over, a big crunch may ensue: a time period when the universe collapses back to its original single point under the forces of gravity. The big bang theory suggests that space is finite as well. Real space would end where the universe ends. At the smallest levels of physics, quantum theory similarly argues that there is no infinity to matter: quantum physics says it is not possible to think of infinitely small measurements. As we can see, empirical science offers a few options concerning the infinity of time and space. Each is based on a different vision of the physical world.

The metaphysical view of infinity might have been the first one that human beings grappled with, and always in symbolic language. In metaphysics, the symbol of the infinite stands for what is absolute, whereas the symbol for the finite means the relative—or what is dependent on or contrary to the absolute. Typically, metaphysics views human existence as a finite reality in relationship to an absolute reality. This also puts limits on human knowledge, including logic, mathematics, and empirical science.

Despite this contingent nature of the human being, a kind of absolute consistency still controls how our minds approach logic, mathematics, and science. Our thoughts cannot cease to be consistent. Our minds cannot affirm A and not A at the same time. We cannot state that reality is logically contradictory and continue to think logically. Science is not possible if our thought ceases to be consistent.

The acceptance of the absoluteness of the consistency of thought is previous to the mere act of thinking. Yes, our thoughts harbor many inconsistencies. We often think of something in one way and then think the contrary five minutes later. However, we know that we changed our thinking in the past five minutes because we have an underlying consistency in the mind. In this argument that the mind has a core element that is absolute we also find the reason that human beings can communicate: they share this consistency in thought.

Now we arrive at one of the great metaphysical questions: Does that mind have this consistency because of an absolute in the universe, or simply because of the powers of the independent human mind based, for example, on the evolution of the physical brain? If we choose a higher absolute in the universe, then we opt for the metaphysical infinite, typically given the name *God*. If we say the apparent consistency of the mind arises from its own finite limitation, then we do not need an absolute to explain our world. This debate, or puzzle, is the basis for the famous ontological argument for the existence of God, a surpassing and infinite absolute.

The Ontological Argument

As I hope to show, the ontological argument is relevant to our discussion about mathematics and language because the argument lies between metaphysics and mathematics. The historical ontological argument has typically employed some of the logical tools that underlie mathematical theory.

Before we go further, we should define our term. *Ontology* means existence. In the eighteenth century, the logical argument for God's existence as a necessary absolute entered Western philosophical writings under the name of the ontological argument, or the ontological proof of God. While some Greco-Roman writings contain a simple form of the argument, it was formulated most precisely by Islamic and Christian thinkers in the Middle Ages.

Any type of reasoning which concludes that God's existence is logical includes our metaphysical intuitions, just as logic and mathematics begin with intuitions. In philosophy, moreover, the ontological argument is unique for how it tries to show that the existence of God is a reasonable conclusion. Its core is the intellectual intuition that it is logically necessary for there to be a reality that is not limited by finiteness. Interestingly, this intuition may exist apart from religion, for as we will see, some mathematicians argue for an infinite absolute. Obviously, though, the ontological argument is a mainstay for religious belief and theology.

But even in religion, different schools of thought have found the ontological argument either helpful or not. For example, the argument—first elaborated by the Muslim philosopher Avicenna (or Ibn Sīnā) of Persia (980–1037) and the Catholic theologian Anselm of Canterbury (1033–1109)—was also accepted in one form or another by René Descartes, Baruch Spinoza, Gottfried Leibniz, and later the mathematician Kurt Gödel. However, it was also rejected by so great a Catholic philosopher as Thomas Aquinas, and then again by Immanuel Kant, both of whom nevertheless did not deny the existence of a supreme being.

The work of Avicenna and Anselm offers a simple basis to follow the historical legacy of the argument. In *The Book of Healing*, Avicenna made the important distinction between the essence (*mahiat*) and the existence (*wujud*) of beings. He then postulated the necessity of a highest being whose existence is necessary (*Wajib al-Wujud*). Anselm is perhaps the most famous Christian thinker to build upon a similar kind of logic. He said that if a being "than which nothing greater can be conceived" did not exist in reality as well as in the mind, then it could not be the greatest being. Therefore, God must exist, for otherwise logic itself is rendered absurd. As Anselm said in his small work *Proslogion*, speaking in the devotional manner of all his writings, "We believe that You are something than which nothing greater can be thought. . . . And surely that-than-which-nothing-greater-can-be-thought cannot exist in the mind alone. For if it exists solely in the mind, it can be thought to exist in reality also, which is greater." Descartes, Spinoza, and Leibniz all built upon this basic kind of logic; they tended also to argue that since a necessary quality of absolute perfection is to have existence, then God must have existence. In the twentieth century, Gödel revived the validity of Anselm's logic by turning it into a purely formalized argument that was valid in modal logic, a system of logic that looks at what is possible, what is impossible, and what is necessary.

This relationship of the ontological argument to logic and mathematics remains a fascinating topic for us today, including for some mathematicians. Most important, it shows us the difference between the use of mathematical and logical signs and the use of metaphysical and religious symbols. The ontological argument cannot be made, so to speak, in the formal signs of mathematics alone, for it also requires the addition of metaphysical symbols.

The main criticism of the ontological argument is that it tries to prove the existence of metaphysical reallity (the greatest that can be thought) as it were a mathematical or physical reality to prove. For example, the critic would also say that while the ontological argument uses formal logic, it does not prove the existence of God in the same way that a mathematical proof verifies the existence of a mathematical object, such as the fact that, given a prime number, there is always a greater prime number. In verifying a mathematical object, such a proof begins with a mathematical intuition. In the case of a proof for the necessity of God's existence as an absolute, the formal logic is based on a distinct metaphysical intuition—the presumption, that is, that such a universe with an absolute seems appropriate. For the critic, there is no evidence for this beginning intuition.

As noted earlier, the two well-known opponents of the ontological argument were Aquinas and Kant. Both of them argued that from the basic intuition about existence, it is not possible to argue that something real must necessarily exist as well. Aquinas and Kant, in other words, questioned whether we can leap from what is logical to what is real. They believed the flaw of the argument is that it confuses the formal semantics of mental objects with the real semantics of objects in the real world.

For our purposes, the key question is one of defining existence— something that mathematics, empirical science, and metaphysics all try to do as a beginning premise, very much as an intuition of what is logical and appropriate. How does mathematics define existence? When we state that a solution exists to the equation $x^2 = 4$, we are saying that mathematical objects exist. In this case, the objects are the numbers 2 and −2, which are the solution to this equation. The numbers 2 and −2 can be formalized and represented as formal sets.

Empirical science has another kind of definition of existence. A compelling example of how science affirms existence came on August 24, 2006, when the International Astronomical Union (IAU) ruled that Pluto can no longer be considered a "planet" and the number of planets in our solar system was thus reduced from nine to eight. This kind of definition relates to the existence of empirical objects: the planets are physical places to which our spaceships could travel. But they are also objects that we place in categories of existence, such as planet or nonplanet (as is the case now with Pluto).

Metaphysics has yet another definition of existence: the global question of how contingent, or finite, beings have a relationship to being itself—that is, a necessary supreme being. Spinoza posed this question in his famous philosophical work, *The Ethics*. He said that God necessarily exists because God's essence is to exist. Spinoza was making a metaphysical statement about the existence and essence of God. For Spinoza the existence of God coincides with the essence of God, unlike the existence of all the other beings, which are distinct from their essence. This identification of essence and existence in God and their separation in other beings is valid insofar as it is confirmed by a metaphysical intuition or experience.

We refer to Spinoza's metaphysics as *pantheism*, since he closely ties the physical world to God's existence. Thinkers such as Anselm and Descartes, however, came down on the side of traditional theism by saying that while God's essence allows other beings to exist, God's existence is separate from that of all other beings. Despite these differences, all of the ontological arguments agree on the metaphysical intuition that there must be a being without limits. To quote Anselm again, this is a being such that we cannot think of any greater being. Based on this premise, all of the ontological arguments use valid formal logic to show that the existence of God is reasonable.

Formalization of Ontological Argument in Predicate Logic

Now let's look at how such an argument, which might seem purely religious, can actually be made using the signs of logic and mathematics. For readers who are unfamiliar with the way logic is written as a string of signs, it will be enough to see that verbal statements about the world can be translated into these strings of notation. By looking at this use of signs closely, we can better understand the difference between formal signs and metaphysical symbols.

We start with a simplified formalization of the ontological argument (OA), which is based on a metaphysical sentence about the existence of something without limits.

OA:

(1)   God is the greatest being we can think of.

(2)   If God does not exist necessarily in reality, then we can think of an existing being greater than God.

∴   Therefore God exists in reality.

Taking the propositions (1) and (2) as premises, we can formalize the argumentation OA in classical predicate logic.[1] To do so, we first define the predicates GR(x,y) and E(x) as follows:

GR(x,y)   :↔   We can think that x is greater than y.

E(x)   :↔   x exists in reality

We can also use the letter g in order to formally designate God:

g   :↔   God

Using these signs, we formalize OA:

(1) $\neg \exists\, x\, GR(x,g)$

(2) $\neg E(g) \rightarrow \exists \times GR(x,g)$

∴ $E(g)$

where we can read the premises (1) and (2) and the conclusion ∴ as follows:

$\neg \exists \times GR(x,g)$

There does not exist an element of our domain of discourse x such that we can think that x is greater than God.

(2) $\neg E(g) \rightarrow \exists \times GR(x,g)$

If God does not exist in reality, then there is an element x of our domain of discourse such that we can think that x is greater than God.

∴ $E(g)$

God exists in reality.

In the formalization of the ontological argument, we have used two different signs, the logical sign ∃ and the predicate E, in order to denote the real existence of the same element x. The use of the logical sign ∃ in formula (2) indicates the existence of an arbitrary element x of our domain of the discourse to which we attribute the predicate GR(x,g). The use of the predicate E in the subformula ¬E(g) indicates the real nonexistence of g.

It is easy to verify that, by applying the rules of deduction of classical logic to the formal propositions (1) and (2), the conclusion (∴) follows.

In fact, classical logic admits the principle of *tertio excluso* whereby, given a predicate E and an object g, or E(g) or ¬E(g), one of the two is true. If ¬E(g) is true, by the conditional proposition (2) we would obtain that ∃ x GR(x,g) is true. This would make the principle of noncontradiction false as both ∃ x GR(x,g) by (2) and ¬∃ x GR(x,g) by (1) are true. Therefore, we conclude that if (1) and (2) are true, E(g) must be true.

The formalization of the ontological argument shows that we can construct a formally correct ontological argument and conclude that God exists, in the case that the two premises (1) and (2) are true. However, the problem stated by Aquinas and by Kant still is not addressed directly in the above solution. Aquinas and Kant argue that we cannot confuse existence as thought with existence in reality. They would argue that in premise (2) we are confusing the logical sign ∃, which refers to formal existence, with the predicate E, which refers to necessary real existence. That is to say, we are confusing the real world with the mental world.

How does the evidence that was so persuasive to Avicenna, Anselm, and Descartes in the ontological argument relate to the kind of evidence demanded by empirical science? Obviously, empirical science required confirmation by the physical senses. One good example is the geometry of the German mathematician Hermann Minkowski, who offered propositions about Einstein's theory of relativity. The Minkowski propositions must be tested against physical observation and measurement. A metaphysical proposition is different: it must be confirmed by the intuition that there must be such a thing as unlimited existence.

Supporters of the ontological argument as valid, therefore, would accept the formal proposition $\neg E(g) \rightarrow \exists\, x\; GR(x,g)$ as an expression of a metaphysical perception, intuition, or evidence. Again, the metaphysical intuition is not apprehended like the empirical object of a planet or the bending of light. It is also not like the purely formal perception behind the existence of a mathematical object. Metaphysical proof has its own way of being thought and felt. The use of the logical sign $\exists$ in formula (2) indicates the existence of an arbitrary element x of our domain of the discourse to which we attribute the predicate $GR(x,g)$. This element x can exist only in our mind. It is a metaphysical option to give it real existence. The use of the predicate E in the subformula $\neg E(g)$ indicates the real existence of g.

Anselm and other defenders of the ontological argument reasoned correctly from a formal point of view. Their conclusion is valid in the case that the following metaphysical premise is valid: "If God does not exist in reality, then there is an element x of our domain of discourse such that we can think that x is greater than God." But although that premise can be accepted as metaphysical evidence, it is not empirically or mathematically true.

We can also look at this from the viewpoint of predicates. For example, $GR(x,g)$ is a predicate with two arguments that we can interpret in a model of reality as saying: the element x of the domain of discourse is greater than the element g. On the other hand, we could say that $E(x)$ is a predicate with an argument that we can interpret as saying: the element x of the domain of discourse exists in reality.[2] The fact that the element g does not have the property E (of existing in reality) does not imply that x is greater than g. In other words, $\neg E(g)$ can be true and $GR(x,g)$ false.

These kinds of formal arguments can become more and more sophisticated, but they invariably move us further away from reality. In summary, the fact that we can think of an object that is bigger than God does not mean that such an object really exists. It may be an object simply created by the mind—a figment of the imagination. Once again we see the importance of choosing a beginning premise for our logical systems. Different communities choose different premises, and therefore some agree with the ontological proof, and others do not.

Abrahamic Religions and Metaphysics

Our discussion of metaphysics suggests that it has a natural fit with the monotheistic religions, which we also call the Abrahamic faiths, since they all trace themselves to the early monotheism of Abraham: Judaism, Christianity, and Islam. All of these traditions have relied upon tools of logic and metaphysics to show the reasonableness of belief in the existence of the absolute. However, they also go a step further than relying on a kind of logical argument based on the intuition that an absolute must exist. They also teach the need of having faith in a transcendent God that is beyond logic and mundane evidence.

This faith requires the belief that God, acting independently of the world, has revealed his existence to the world, an existence that the human mind cannot entirely apprehend. A kind of revelation given to the world must also be at play, according to this Abrahamic tradition. Often this metaphysical perception of God leads individuals to feel a personal mission they must undertake to the world. Many such testimonies are present in history. We can think, for example, of the prayers of the Englishman John Henry Newman (1801–1890), who explored both the Anglican and Roman Catholic traditions.

"God has created me to do Him some definite service," wrote Newman, who became a Catholic cardinal and university educator. "He has committed some work to me which He has not committed to another." In my own specific tradition, the founder of the Jesuits, Ignatius of Loyola (1491–1556), also offered an interpretation of human purposes in the world in the opening section of his *Spiritual Exercises*, titled "Principle and Foundation":

> The human person is created to praise, reverence and serve God Our Lord, and by so doing to save his or her soul. The other things on the face of the earth are created for human beings in order to help them pursue the end for which they were created. It follows from this that one must use created things in so far as they help towards one's end, and free oneself from them in so far as they are obstacles to one's end. To do this we need to make ourselves indifferent to all created things, provided the matter is subject to our free choice and there is no prohibition.

The language of such testimonies cannot be mathematical or scientific. It must be a language of theological symbols. Those symbols, in turn, are moral and ethical. They are more than a narrative, for they also speak of the meaning of human action in the world and a daily relationship of the human being to the absolute. This language—often the passionate language of faith and belief—is about attitudes and relationships, about such experiences as trust, mercy, and forgiveness.

Summarizing the Three Languages

At this point in our comparison of the languages of logic/mathematics, empirical science, and metaphysics, it could easily seem that we are looking at three entirely different worlds. The gap may seem especially large between metaphysical symbols and the signs of math and science. However, the three languages can also be seen as complementary because, by applying them all, we speak to the full reality of the human experience. None of these languages can be excluded. As we saw earlier, while logic may not need religion, religion does need to harmonize itself with logic and science.

The key difference between these languages relates to the use of signs and symbols. In our day-to-day language, we don't really see any difference between the words "sign" and "symbol," and indeed the distinction is of more interest to academics. But the distinction is helpful to everyone if we want to understand the interaction of our three kinds of languages: formal, representational, and metaphysical.

In our use of logic, formal signs make it possible to write propositions that express evidence. For example, the set of formal signs $\{\neg, A, \wedge, (,)\}$ makes it possible to write the proposition $\neg(A \wedge \neg A)$. This is the basic principle of noncontradiction. Furthermore, the formal signs of arithmetic allow us to express mathematical intuitions regarding numbers and certain numerical operations. For example, the set of formal signs $\{2, 4, +, =\}$ makes it possible to write the arithmetical proposition $2 + 2 = 4$.

Our second kind of language uses the representational signs of physics and the empirical sciences. This language points to observable realities such as the mass, speed, or energy of physical bodies. The set of representational signs $\{E, m, c\}$ gives us the ability to talk about energy (E), mass (m), and the speed of light (c) in the useful equation $E = mc2$, which expresses a property of the physical theory of relativity.

Finally, because our minds are able to logically ask ultimate questions, we need a metaphysical language to speak of what mathematics and science cannot touch. These metaphysical questions and answers need a language of meaningful symbols that are neither simply formal nor empirical. The metaphysical meaning of symbols expresses personal evidence about the ultimate principles of reality. The people who specialize in logic, mathematics, and science are all members of communities, and these groupings influence, even bias, the assumptions made in using these generally objective types of languages. This social context is even more important in metaphysical language, however, for only by having a distinct community—a history, tradition, and core values—can this language retain its powerful meaning.

Notes

1. Readers who are interested can find a brief presentation of the syntax of first-order predicate logic in appendix 3 of my book, *Mathematics and Religion* (West Conshohocken, PA: Templeton Press, 2010).
2. Readers can find more details on the semantics of first-order predicate logic in ibid., appendix 4.

Noreen Herzfeld
Technology and Religion

 "TECHNOLOGY IS HERE TO STAY," Herzfeld says in her book *Technology and Religion.* Even so, it is part of human wisdom that we "should not blindly espouse every technology that comes down the pike." Herzfeld shows how modern technology holds promise and peril in many areas. These include the bioethics of birth and death, pharmaceuticals, bionic devices, genetic engineering, cyberspace, altering matter, and energy use. In promising benefits, each area poses practical, ethical, and often theological questions for society. Technology and its economic force come down the pike so rapidly now that the only brake could be religious values, Herzfeld suggests. These values—which often say "No"—have become everyday occurrences for citizens, who read headlines such as, "Is Google Making Us Stupid?" Is technology taking away privacy, even altering our sense of reality? Does the future face a "technological tyranny," as some predict? Whatever the case, Herzfeld says we must prepare to make prudent "high-tech choices," drawing upon religious traditions, preservers of

time-tested wisdom. Many of our values can be used to address technology. They range from the theological idea of creation to biblical calls for humility and justice. Overall, Herzfeld urges us to think about human "relationship and responsibility" as a guide for living in a technological world.

CHAPTER 8

Between Cyberspace and the New Alchemy

Noreen Herzfeld

AMONG OUR MANY new technologies of "mind and matter," computers and nanotechnology (the smallest scale of chemical engineering) represent two innovations on the cutting edge of our scientific future. Each one presents our value systems with a unique challenge. Computers and the Internet are forcing much of humanity to adapt to cyberspace, willing or not. In the meantime, nanotechnology represents a "new alchemy." It raises the question of how far we will go in altering physical matter itself.

Our theological traditions demand that we grapple with the ways that computers and nanotechnology are redefining our world. In the Gospel of John, Jesus asks his followers to plunge into the world: "I am not asking you to take them out of the world, but I ask you to protect them from the evil one" (John 17:15). Similarly, the prophet Muhammad eschewed monasticism, expecting his followers to live active lives rather than only "sit in the corner of a mosque in expectation of prayers."[1]

In whichever direction we look, the "real" world that religions speak about is being touched by the computer-generated "virtual" world. At the same time, the new alchemy presents us with both promise and peril as we seek to use it to alter the "real" world.

HUMAN AND ARTIFICIAL INTELLIGENCE

The idea of the computer has become so pervasive in our society that both computer scientists and brain researchers use terms such as "hardware" and "software" to speak of our minds and the thoughts they produce. According to recent research, the computer (even more than drugs, perhaps) is altering both our mental hardware (the brain's neuronal structures) and our mental software (our ways of perceiving). Tools such as Facebook and smart phones reshape our social relationships. Just as significantly, artificial intelligence (AI) research has given us new ways to think about the human mind and intelligence.

The computer has raised once again what were once considered theological questions. What is the mind? Where is consciousness? Do we have souls, and if so, how do they relate to the mind or the body? These questions underlie many of the ethical issues that bedevil both scientists and politicians in the twenty-first century. In the real world, controversies surrounding abortion and stem-cell research are rooted in varying views of when a new human soul comes into being. Euthanasia asks a similar question in reverse, namely, when does the soul depart in death? If our mental reality is, as the computer scientists and brain researchers suggest, analogous to the hardware and software of a computer, we must now also ask whether a computer itself could have a soul.

The key to this puzzle is in how we define *intelligence* and *soul*. These two concepts are not as easy to define as we might expect. Most of us have an intuitive understanding of what we and others mean when using those terms, but would find it difficult to give a satisfactory definition of either. Yet a clearer understanding of both is crucial if we are to sort through the controversies mentioned above. Our approaches to designing artificially intelligent machines give us one avenue toward understanding the different aspects of intelligence and how they relate to the vexing question of the soul.

Three Approaches to Intelligence

In designing an intelligent computer, the first approach has been to define *intelligence* as the ability to solve problems. This definition fits our intuitive notion of intelligence, based as it is on the model of activities that we consider indicative of highly intelligent people, such as the ability to play chess or solve complicated equations in mathematics or physics. Computer scientists in the 1960s and 1970s took mathematics as a model (after all, most early computer scientists were mathematicians). Just as the field of geometry is built from a finite set of axioms and primitive objects such as points and lines, so early AI researchers, following rationalist philosophers such as Ludwig Wittgenstein and Alfred North Whitehead, presumed that human thought could be represented as a set of basic facts. These facts would then be combined, according to set rules, into more complex ideas. This approach to AI has been called "symbolic AI." It assumes that thinking is basically an internal process of symbol manipulation.

Symbolic AI met with immediate success in areas where problems can be described using a limited set of objects or concepts that operate in a highly rule-based manner. Game playing is an obvious example. The game of chess takes place in a world in which the only objects are the thirty-two pieces moving on a eight-by-eight board, and these objects are moved according to a limited number of rules. Other successes for symbolic AI occurred rapidly in similarly restricted domains, such as chemical analysis, medical diagnosis, and mathematics. These early successes led to a number of remarkably optimistic predictions of the prospects for symbolic AI.

Symbolic AI faltered, however, not on difficult problems like passing a calculus exam but on the easy things a child can do, such as recognizing a face in various settings, riding a bicycle, or understanding a simple story. One problem with symbolic programs is that they tend to break down at the edges; in other words, they cannot function outside or near the edge of their domain of expertise since they lack knowledge outside of that domain, knowledge that we think of as common sense.² Humans make use of millions of bits of knowledge, both consciously and subconsciously. Often we do not know what bits of knowledge or intuition we have brought to bear on a problem in our subconscious minds. Symbolic AI programs also lack flexibility. In 1997 the computer Deep Blue beat then–reigning world chess champion Gary Kasparov. In the ten years since, Deep Blue's successors continue to play chess—and only chess. Kasparov, on the other hand, has become a politician and presidential candidate in Russia. Intelligence seems to be a quality that is larger than can be captured in any symbolic system.

A second way to look at intelligence is as the ability to act within an environment. This means that, first of all, intelligence is embodied. Of course, any intelligent agent would be embodied in some way. Deep Blue did not have what we would think of as a body; it could not pick up the chess pieces and physically move them. However, the program was embodied in a bank of supercomputers. So the question is not whether intelligence requires a physical body but what kind of body. Does a humanlike intelligence require a humanlike body?

Our bodies determine much of the nature of our interaction with the world around us. Our perception is limited by our physical abilities. For example, we think of location in two-dimensional terms because we walk rather than fly. We evaluate a situation primarily by sight and sound; anyone who has ever walked a dog knows that the animal evaluates with nose to the ground, receiving a whole different data set from what we see. In a similar fashion, our physical makeup determines how we act within the world. Our opposable thumbs, combined with the softness and pliability of our skin and underlying muscles, allow us to grasp and manipulate objects easily. When we ride a bicycle, we need not calculate equations of force, trajectory, and balance; our muscles, nerves, and inner ear do the work for us. In fact, should we begin to make such calculations consciously, we are likely to fall off! Most athletes know that they perform at their best when their minds are in a meditative, rather

than a discursive, mode. As one of the characters says in Fred-
erich Nietzsche's *Thus Spake Zarathustra*, "Behind your thoughts
and feelings, my brother, there stands a mighty ruler, an unknown
sage—whose name is self. In your body he dwells; he is your body.
There is more reason in your body than in your best wisdom."[3]

In human history, the tools we make reflect our embodied state
of intelligence. Anthropologists pour over ancient blades, weapons,
and agricultural devices to understand past modes of thinking. The
extended physical experiences of our forebears shaped the ordinary things we use today. Philosopher John Haugeland explains:

> Think how much "knowledge" is contained in the tradi-
> tional shape and heft of a hammer, as well as in the mus-
> cles and reflexes acquired in learning to use it—though
> ... no one need ever have thought of it. Multiply that by
> our food and hygiene practices, our manner of dress, the
> layout of buildings, cities, and farms. To be sure, some of
> this was explicitly figured out, at least once upon a time;
> but a lot of it wasn't—it just evolved that way (because
> it worked). Yet a great deal, perhaps even the bulk, of
> the basic expertise that makes human intelligence what
> it is, is maintained and brought to bear in these "physi-
> cal" structures. It is neither stored nor used inside the
> head of anyone—it's in their bodies and, even more, out
> there in the world.[4]

Our designs and behaviors arise through and out of interaction
with the environment. And the type and extent of this interaction
are determined by the body.

At first glance, it seems that our ability to make plans could break
this rule of embodiment. Certainly, we can plan a vacation, draw
up a new living-room design, or arrive at a political opinion by "the
mind alone." Neuroscience, however, reveals a more complex situation. In 1983 Benjamin Libet conducted a series of experiments in

which the subject was asked to make the simple decision to move a finger and to record the moment this decision was made. Sensors also recorded the nerve impulse from brain to finger and found that the impulse was on its way roughly half a second before the subjects consciously registered that they were going to move their fingers. Thus, it seems that the choice preceded conscious reasoning. The subconscious mind and the body had things under way before the conscious introspective mind knew about it.[5] Anyone who has driven a car while talking on a cell phone, or while thinking over the day's schedule, knows that the subconscious mind and body can keep things in hand, up to a certain point, while the conscious mind works on other things.

The idea that a computer needs a body has always had its niche in the world of artificial intelligence. Almost every artificially intelligent computer that has appeared in the realm of science fiction has been a robot, often with a more or less humanlike body.[6] Rodney Brooks at MIT notes that the basic problems with symbolic AI are rooted in the fact that the problem-solving programs it produces are not situated in the real world. Brooks, and others at a variety of AI labs, have built a series of robots that act within the world on the basis of data acquired through sensors. Brooks began with a series of insects, later moving on to the humanoid robots Cog and Kismet, which acquired some of the rudimentary skills of a baby through interaction with human beings. None of these robots comes close to humanlike intelligence, but some seem to have a niche in their environment. Consider the Roomba, a roboticized vacuum cleaner that navigates around a room looking for dirt, avoids furniture and stairs, and plugs itself in when it needs to be recharged.[7] One might argue that Roomba shows as much intelligence as many animals in its ability to navigate in a local environment, avoid hazards, and forage for sustenance.

During World War II, the British mathematician Alan Turing found a way to unscramble German codes, revealing their battle plans to the Allies. Naturally, this brilliant exploit prompted him to think about the nature of intelligence. He concluded that it is a relational process—which is the third kind of approach seen in artificial intelligence today. We guess at someone's intelligence, in other words, by talking to that person. In his 1950 landmark paper, "Computing Machinery and Intelligence," Turing presented a test for deciding if a computer is intelligent. In this test, an interrogator slips written questions into a room through a slot and, by the written replies that come back, tries to decide whether the denizen in the room is a human or a computer. If the interrogator fails as often as she succeeds in determining which was the human and which the machine, the machine could be considered as having intelligence.[8] Turing predicted that by the year 2000, "It will be possible to programme computers . . . to make them play the imitation game so well that an average interrogator will not have more than a 70 percent chance of making the right identification after five minutes of questioning."[9] This, like most predictions in AI, was overly optimistic. No computer has yet come close to passing the Turing Test.

Turing was not alone in turning to discourse as a hallmark of intelligence. Discourse is unique among human activities in that it subsumes all other activities within itself, at one remove.[10] Terry Winograd and Fernando Flores assert that cognition is dependent upon both language and relationships. Objects we have no words for do not exist for us in the same way as those we name. We make distinctions through language. Without words to describe difference, distinctions cannot long be held in mind nor shared with others. But discourse is essentially a social activity. The act of speaking to another is not simply the passing of information. Winograd and Flores note, "To be human is to be the kind of being that generates commitments, through speaking and listening. Without our ability to create and accept (or decline) commitments we are acting in a less than fully human way, and we are not fully using language."[11] Understanding arises in listening not to the meaning of individual words but to the commitments expressed through dialogue. Thus, understanding is both predicated on and produces social ties.

To navigate the world of relationships, one needs what has recently been termed "emotional intelligence." When viewed superficially, emotions seem to obscure rational thought. However, recent research has shown that emotions, far from getting in the way of thought, are actually necessary for cognition. In *Descartes' Error*, Dr. Antonio Damasio notes that patients who have had a brain injury to the parts of the brain that govern the ability to feel emotions also lose the ability to make effective decisions, even decisions as simple as what to have for lunch. Neurophysiologist J. Z. Young notes that "even the simplest act of comparison involves emotional factors."[12] If we have no fears, no desires, we have no reason to value one choice over another. Harvard psychologist Joshua Greene has used brain-imaging techniques to study moral decision making. He notes that our brain automatically generates a negative emotion whenever we contemplate hurting someone. Greene's data suggest that psychopaths cannot think properly because they lack normal emotional responses. "This lack of emotion is what causes the dangerous behavior."[13]

Can a computer have emotions? As computer scientist Marvin Minsky quipped, the question is not whether a machine can have emotions, but whether machines can be intelligent if they do not have emotions.[14] Damasio's and Greene's data show that the answer is no. But how far have we gotten in programming emotions into a machine? Not very far. MIT scientist Rosalind Picard has shown that a computer can be programmed to recognize emotion in either facial expressions or tone of voice. The cute robots in the MIT lab, Cog and Kismet, express a variety of emotions, like fear, amazement, or pleasure.[15] However, while computers can be programmed to recognize or express emotion, actually feeling emotion requires a level of self-consciousness current machines lack.

Turing, Damasio, Winograd, and Flores all view intelligence as based upon some form of social activity. Though they approach it in different ways, each suggests that the idea of an individual intelligence is meaningless; intelligence has meaning only in encounter. Whether a computer could have the capability of entering into true relationship with human beings remains to be seen.

What about the Soul?

In the seventeenth century, the French philosopher René Descartes famously argued that the soul and body are utterly separate. Today, it might seem that the insights of computer-science brain research—with their emphasis on the hardware of the body—would destroy this "Cartesian dualism" once and for all. Oddly enough, this is not the case. The concept of a self separate from the body has been given a recent boost, precisely by computer technology. Many computer enthusiasts today spend large amounts of time interacting in a bodiless world—a "virtual" world that Descartes could hardly imagine. Activities that once took place in real space now take place in cyberspace. Consider: We communicate via Facebook, text messages, and e-mail; we shop, bank, and do research on the Internet; we amuse ourselves with video games, MP3s, and streamed videos, or as avatars in a second life. We project our minds across vast distances or into fictional realms and have experiences in those places that form us as persons.

This does, of course, have certain advantages. Neal Stephenson, in his novel *Snow Crash*, notes that in cyberspace, "If you're ugly, you can make your avatar beautiful. If you've just gotten out of bed, your avatar can still be wearing beautiful clothes and professionally applied makeup."[16] One can project an image of oneself, and that image is utterly malleable, changed at the flick of a bit.

While the ability to design a body amuses, the greatest seduction of a bodiless existence lies in the fact that our bodies are mortal, subject to sickness, aging, and ultimately death. Computer scientist Ray Kurzweil, in *The Age of Spiritual Machines*, suggests that cyberspace provides a place where we can evade the mortality of the body by downloading our brains into successive generations of computer technology. Kurzweil writes:

> Up until now, our mortality was tied to the longevity of our hardware. When the hardware crashed, that was it. For many of our forebears, the hardware gradually deteriorated before it disintegrated. . . . As we cross the divide to instantiate ourselves into our computational technology, our identity will be based on our evolving mind file. We will be software, not hardware. . . . As software, our mortality will no longer be dependent on the survival of the computing circuitry . . . [as] we periodically port ourselves to the latest, ever more capable "personal" computer. . . . Our immortality will be a matter of being sufficiently careful to make frequent backups.[17]

Kurzweil thinks we might achieve this new platform within the next fifty years. He is not the sole holder of this expectation, though he may be among the more optimistic in his timeline. In *The Physics of Immortality*, physicist Frank Tipler conjectures that the universe will cease to expand and at some point end in a contraction that he calls the *omega point*. Tipler sees this omega point as the coalescence of all information, including the information that has made up every person who ever lived. At such a point, the information making up any given individual could be reinstantiated, resulting in a form of resurrection for that person, though Tipler is vague as to how such a reinstantiation might come about.[18]

Both Kurzweil and Tipler hold a worldview that would seem, at first glance, to be at odds with the reductive physicalist position held by many scientists today. Yet their views actually are quite consistent with this worldview. They suggest that the soul is, first of all, nothing more than the collection of memories, experiences, and thoughts that we hold in the neural connections of our brain. In other words, our soul is information. This is seductive for the computer scientist who sees the world in terms of 0s and 1s. Our soul is the information that emerges from the state of consciousness, a quality only held by matter that has evolved or self-organized into a sufficiently complex system.

Biologist Francis Crick expresses this view well: "You, your joys and your sorrows, your memories and your ambitions, your sense of personal identity and free will, are in fact no more than the behavior of a vast assembly of nerve cells and their associated molecules. . . . You're nothing but a pack of neurons."[19] The "you" that Crick speaks of here is not initially disembodied but arises from the workings of the brain. Without such a material basis, "you" cease to exist. But what we identify as "you" is not the brain itself, but the information stored in that brain. In this view, the soul, as information, though dependent on the body initially, later becomes completely separable from the body.

Here we have a first sense of the concept of soul as that part of the self that transcends our mortality. Is doing so on a different platform, such as a computer, consistent with the Christian understanding of the soul's immortality? Not really. The Nicene Creed states that our resurrection is one "of the body," and Paul makes clear that the resurrected body will be a new and different body than our current one (1 Corinthians 15:50). The problem with computer hardware as the platform for this new body is that it is not a part of a new creation but a continuation in this creation. Donald MacKay notes the difference:

> If the concept of creation is to be thought of by any analogy with creation as we ourselves understand it—as, for example, the creation of a space-time in a novel—then a new creation is not just the running on and on of events later in the original novel: it is a different novel. A new creation is a space-time in its own right. Even a human author can both meaningfully and authoritatively say

that the new novel has some of the same characters in
it as the old. The identity of the individuals in the new
novel is for the novelist to determine. So if there is any
analogy at all with the concept of a new creation by our
divine Creator, what is set before us is the possibility that
in a new creation the Author brings into being, precisely
and identically, some of those whom He came to know
in and through His participation in the old creation.[20]

The physicist and theologian Bob Russell once put it to me this
way: "Immortality does not just mean more time." After all, we all
know—scientists and theologians alike—that the Earth is itself
temporal and finite, that "heaven and earth will pass away" (Mark
13:31).

This fact of human mortality, and the finiteness of the world, has
often been used to explain the source of so many human evils and
missteps. In the twentieth century, the Protestant theologian and
social critic Reinhold Niebuhr wrestled with the concept of human
"sin," especially as it played out in utopian projects of communism
and fascism, and even the attempts of democracies to use force and
still claim sinless innocence. He traces sin to human finitude. The
evils of the world, he goes on, often arise from human attempts to
assert powers that belong only to God. We need to accept the finitude of our bodies and minds:

> Man is ignorant and involved in the limitations of a finite
> mind; but he pretends that he is not limited. He assumes
> that he can gradually transcend finite limitations until
> his mind becomes identical with universal mind. All of
> his intellectual and cultural pursuits, therefore, become
> infected with the sin of pride.[21]

In this post-Cartesian age, we have come to realize that human
beings have a single nature with two inseparable elements: a self-
transcending mind and a finite body. Denying the body has led

to a worldview that denigrates both the natural environment and women. For if we could live in the bits of a computer, of what use is the natural world? If we can replicate ourselves through backup copies, who needs babies or even sexual differentiation? Here I note, however, that, while for them it serves no reasonable purpose, proponents of cybernetic immortality are loath to give up sexual experience itself. Tipler waxes eloquent on the possibility of fulfilling all our sexual desires at his omega point, and Kurzweil is equally enthusiastic about the possibilities of disembodied sexual experience.[22] But these experiences are viewed only in terms of self-gratification, not as true relationship, with all the complexity that that entails.

The relational nature of intelligence suggests that the model of an artificial intelligence that holds a separate identity and acts by itself in the world, as a replacement for human intelligence, is the wrong model. What we truly need are machines that complement what people do, working with human beings to accomplish tasks we cannot do alone. Yet there is one caveat. Human beings are far more flexible than computers. We easily overidentify with and overuse our machines. We all see this in our society's current obsession with quantifiable data. Programmer and commentator Jaron Lanier suggests that, should a computer actually pass the Turing Test, it might not be the case that the computer has become smarter or more human but that our immersion in a computerized world has led humans to become more like machines. Miniature Turing Tests happen whenever we adapt our way of acting or thinking to our software; "we make ourselves stupid in order to make the computer software seem smart."[23] Ethicists Joanna Bryson and Phil Kime have pointed out that our overidentification with computers has led "to an undervaluing of the emotional and aesthetic in our society. Consequences include an unhealthy neglect and denial of emotional experiences."[24] The moral of the story? We can protect the emotional richness of life if we see intelligence as a complex phenomenon, rooted firmly in our bodies as well as minds.

The mystical side of many religious traditions has offered another insight that may help us navigate our world of computers. The mystics have long suggested that we are deeply connected to one another. There seems to be a technological corollary to this idea of connectedness as well. According to our current understanding of intelligence, both human and artificial, intelligence, consciousness, and yes, probably the soul as well, are meaningless outside of the context of the human being in a web of relationships with other humans and with the environment.

Much of that environment is nature itself, held together by the properties of chemistry. How far do we want to go in manipulating the chemical properties of nature at its smallest levels, and to what end? The question is not new, and has been pondered over the centuries by religious and scientific thinkers.

THE ETHICS OF THE NEW ALCHEMY

A few centuries ago, the German theologian Martin Luther raised a technological topic during his famous "table talks," and later in London, the scientist Isaac Newton pondered the same subject. They called it "alchemy," the power to change matter itself. "The science of alchemy I like very well," Luther said, probably over a pint of Wittenberg ale. "I like it not only for the profits it brings in melting metals, in decocting, preparing, and extracting and distilling herbs, roots; I like it also for the sake of the allegory and secret signification."[25] The dream, of course, was to change base metals into precious gold. In many ways, that dream has not died. On the border of France and Switzerland, the largest machine in the world—the seventeen-mile circular Large Hadron Collider—despite current difficulties, is designed to crash two particle beams into each other at the highest rates of energy ever achieved. The goal is to "decoct" matter itself into its smallest parts and, in the process, to discover the "secrets" of the universe.

In our long history of inventing technologies of matter, the goal has always been to control the natural world around us. As the French philosopher Jacques Ellul, who shaped much of the debate over modern technology, remarks, technology can isolate us from nature; today we have a wall of televisions, computers, smartphones, cars, and other machines between us and the great outdoors. Nevertheless, technology is essentially about control of nature. It is about taking hold of matter and wrenching from it what we want.[26]

Our modern approach to the technologies of matter has one significant difference from that of our forebears, according to historians of science and culture. In the past, the natural philosophers used to see life as embedded within material objects, a view called "animism" or "vitalism." Religious traditions such as Christianity have espoused the idea of stewardship, of care for these natural things as if they were living partners in the Creation. However, the modern mentality sees objects as nothing more than chunks of matter to be used by human ingenuity. In fact, the biblical idea of "dominion" can easily be read in this utilitarian way. In 1967 historian Lynn White stirred a great controversy by pointing this out in a widely read article in *Science* magazine:

> In Antiquity every tree, every spring, every stream, every hill had its own genius loci, its guardian spirit. These spirits were accessible to men, but were very unlike men; centaurs, fauns, and mermaids show their ambivalence. Before one cut a tree, mined a mountain, or dammed a brook, it was important to placate the spirit in charge of that particular situation, and to keep it placated. By destroying pagan animism, Christianity made it possible to exploit nature in a mood of indifference to the feelings of natural objects.[27]

Few today worry about the feelings of natural objects. The ecological crisis of today, however, may be changing our indifference to the natural world. Our technologies have become more power-ful, and we are seeing more clearly their side effects, both on nature and society. This is a good time to recall the questions our religious traditions have provided for assessing our technologies: First, does the technology provide tangible benefits to the community or individuals within that community? Second, does the technology change the relationship of the individual to the community? Third, does the technology change the nature of the community? When it comes to technologies of matter—which seek to change nature itself—the community is large. We are challenged to consider not only how revolutions in technology affect the poor and marginal in the world, but even how they affect other species and the very planet itself.

A case in point is the rise of nanotechnology.

NANOTECHNOLOGY

As we know, the scientists and philosophers of Luther's and New-ton's eras never did transmute base metal into gold, nor has any-one since. But we still continue the quest for what they called the "philosopher's stone," which stands for a material, process, or elixir with extraordinary powers to alter nature. To its credit, alchemy did produce a wide array of useful chemicals and pharmaceuticals. The new field of nanotechnology has likewise produced a variety of new and improved products and promises more. But the similarity between alchemy and nanotechnology does not stop there. Nan-otechnology pursues its own philosopher's stone. Its enthusiasts promise us the power to change one form of matter into another, bringing wealth, restoration of the body—and even eternal life.

Materials Technology: Tennis Balls and Sunscreens
Nanotechnology is the design and production of devices or materi-als on the scale of 1 to 100 nanometers (nm), where a nanometer is one-billionth of a meter. To place this scale in perspective, human

hair ranges in width from 25,000 to 50,000 nm, a DNA molecule is 2.5 nm, and most proteins, 1 to 15 nm. At the nanometer scale, one is essentially working at the molecular or atomic level. Techniques in nanotechnology could ultimately allow us to design anything at that level, attaching whatever molecules we like so long as their attachment is consistent with the laws of physics. This would allow the production of materials that are not found in nature, materials that are lighter, stronger, and more precisely tailored than we can currently produce. It could also give us machines small enough to navigate the human bloodstream.

Nanotechnology represents the newest frontier of both materials science and medicine, on which both businesses and governments are anxious to take the lead. It is less a field in itself than a subfield in each of the fields of physics, chemistry, engineering, computer science, biology, and genetics. The wide diversity of applications in nanotechnology allowed the field to sweep the Nobel Prize competition in 2007, in which the prizes in medicine, physics, and chemistry all involved research at the molecular or nanoscale level. As the newest frontier in both science and engineering, nanotechnology has garnered extreme private and public interest and support. The National Nanotechnology Initiative in the United States has gone from a budget of approximately $500 million at its founding in 2001 to approximately $1.3 billion of governmental funding in 2006; internationally, nanotechnology research dollars totaled $9.6 billion in 2006. The total value of all products incorporating nanotechnology was estimated to be $50 billion in 2006 and is projected to reach $2.6 trillion by 2014.[28]

Ideally, nanotechnology would let us assemble any substance we wish, snapping together the basic molecules like so many Lego blocks. This would require tools that allow us to manipulate individual atoms and molecules. The first such tool was the scanning tunneling microscope, for which Gerd Binig and Heinrich Rohrer received the Nobel Prize in physics in 1981. This microscope and subsequent similar tools allow researchers to "see" at the molecular level, to manipulate molecules, albeit so far only in a two-dimensional way, and to see how nanoscale products interact with cells and other substances. To manipulate molecules on a three-dimensional level, the optimal such tool would be a molecular robot, one that is itself molecular in scale, programmed to manipulate objects at its own level. Physicist Richard Feynman introduced the idea of nanobots in 1959: "I want to build a billion tiny factories, models of each other, which are manufacturing simultaneously. . . . The principles of physics, as far as I can see, do not speak against the possibility of maneuvering things atom by atom. It is not an attempt to violate any laws; it is something, in principle, that can be done; but in practice, it has not been done because we are too big."[29]

The machines envisioned by Feynman do not yet exist. Current nanomaterials have been produced in a bottom-up fashion by relying on natural chemical processes to do the assembly for us. These nanosubstances spontaneously self-assemble when triggered by some change in their environment, such as acidity, temperature, or the application of an electric charge.[30] One of the most celebrated nanomaterials, carbon nanospheres (also known as buckminster-fullerenes or buckyballs), is constructed through molecular self-assembly, in which the carbon molecules are induced to arrange themselves automatically into either a soccer-ball-like configuration or a very thin cylinder when an electric current is passed between two graphite electrodes.[31] Carbon nanotubes are highly conductive of both electricity and heat and are one hundred times stronger than steel, giving them excellent potential in the design of tiny semiconducting circuits or energy-storing devices. They are currently mixed with other materials in the production of super-strong tennis rackets, lightweight bulletproof vests, and computer screens.[32] Researchers at the University of California have used carbon nanotubes to construct nanoscale radio components.

A second way to produce materials with features at the nanometer scale is from the top down. In this approach, one begins with a larger piece of material and uses photolithography or etching to add nanoscale features. An example of the top-down approach is the production of silicon microchips with nanoscale-sized circuits etched into them. Miniaturized hard disks, with magnetization at a nanoscale level, are used for data storage in almost all personal computers and handheld devices, minimizing the space needed for the storage of large amounts of data.[33]

A variety of inert materials have been developed at the molecular level. These include metals, ceramics, coatings, polymers, colloids, and aerosols that have been produced through structured chemical processes. There are already more than six hundred new products on the market that involve nanomaterials, ranging from sunscreens to catalytic converters, stain-resistant pants to tennis balls. As a branch of materials science, nanotechnology already provides us with improved products and tools that, while not life-changing, are certainly life-enhancing. New water-resistant fabrics are woven with pores at the nanometer level. These allow water vapor to escape yet are too small for water droplets to penetrate. A similar internal coating keeps tennis balls from losing air. This technology might soon be applied to car tires, resulting in an improvement in tire inflation that could lead to a significant savings in fossil fuels and reduced pollution.

One of the largest current markets for nanomaterials is the cosmetics market. Titanium dioxide and zinc oxide are found in a variety of sunscreens and cosmetics. The particles used in their formulation are too small to reflect visible light and, thus, are invisible while being large enough to block the shorter wavelengths of ultraviolet light. Other cosmetics tout the virtue of particles small enough to penetrate the cells of the outer layer of skin.

Pharmaceuticals also stand to gain from nanotechnology in a big way. Nanosilver appears widely in a variety of products, ranging from food storage containers to shoe liners, due to its antimicrobial properties. Even the ancient Greeks and Romans used silver as a topical antibiotic. Silver is toxic to germs, even viruses, because it bonds to various parts of the organism. Yet, in minute concentrations, it does not harm human cells. In 2006 the government of Hong Kong sprayed a nanosilver coating on the handrails of the city's subway system in hopes of slowing the spread of avian flu.

Research has recently moved from the realm of inert materials into the realm of structures that change properties while in use. This is particularly important in medical uses of nanotechnology. There is much current research into nanoscale polymers for drug capsules, which release drugs only under certain conditions, such as the presence of acid or heat. These could be used to deliver drugs to specific locations or to target particular tissues, such as tumors. Doctors at the University of Texas have been experimenting with gold-coated glass nanoshells, which enter tumors through the blood vessels that feed them and are then heated with a laser to burn away the gold, and with it, the malignancy.[34] Pharmaceutical companies are also investigating the reformulation of several drugs themselves into nanoscale particles, which might be better absorbed by the body. Nanoscale diagnostic chemicals, such as those used for contrast in magnetic resonance (MR) imagery, show promise for improved targeting of particular organs or cells; other forms are being developed that will last longer in the body, allowing for MR imaging of arteries and veins.[35]

The primary short-term medicinal use of nanotechnology may well be a fairly simple one—the development of inexpensive water filters. Diarrhea is the leading cause of death among children in the developing world. According to the World Health Organization, waterborne diseases cause the death of more than 2.2 million people per year.[36] Carbon nanotubes can be woven into a filter with nanosized pores that allow water particles to pass through but block larger contaminants, such as chemical pollutants and bacteria. Such filters could be used not only for purification but for desalinization, and they accommodate a fairly rapid flow, since nanotubes are straight as opposed to conventional fibers, whose bends and convolutions impede water flow. A second approach to water purification involves the use of nanoparticles to absorb arsenic and other large contaminants.

The extremely small size of nanoparticles, and their unpredictable nature, makes nanotechnology hard to regulate. Whereas over $1 billion has been allocated to nanotechnology research in the United States each of the last few years, only $38 million of that goes to assessing the environmental or human risks, according to David Rejeski, director of the Project on Emerging Nanotechnology at the Woodrow Wilson Center. There is neither a federal agency nor any statutes to govern nanomaterials. In the United States, they might be regulated by the Occupational Safety and Health Administration during production, then by the Food and Drug Administration during use, falling to the Environmental Protection Agency when they enter the disposal segment of their life cycle. Regulations, moreover, are hard to draw up for such tiny materials, which are hard to monitor outside the laboratory.[37]

Nano-unpredictability extends to health and chemical safety. If inhaled, the particles are more likely than other substances to breach the blood-brain barrier. Research has shown that some nanoparticles, including carbon nanotubes, can result in the kind of protein fibrillation that is characteristic of Alzheimer's and Parkinson's disease.[38] By their nature, nanoparticles are chemically unpredictable, having a "high potential for being surprised," says Rejeski.[39]

At the scale of nanoparticles, some elements behave in new ways, causing unusual "quantum" effects. For example, metals such as copper or zinc become transparent. Other metals become combustible. At the nanoscale, gold turns to liquid at room temperature and becomes a catalyst, rather than an inert substance. Silver becomes antimicrobial. For now, however, nanotechnology seems to be under the control of the usual safety procedures. So far, it has not produced anything that utterly changes our world. But to hear Eric Drexler, that time may come soon.

The Philosopher's Stone

Drexler, one of the earliest visionaries of nanotechnology, explored the notion of molecular manufacturing in his 1986 book *Engines of Creation*. He also cofounded the Foresight Institute, which is pushing for guidelines to avoid mishaps with nanotechnology. Drexler worries about what he calls the "gray goo problem," the risk of fundamentally altering the chemistry of the world. Scientist Bill Joy notes, "Gray goo would surely be a depressing ending to our human adventure on Earth, far worse than mere fire or ice, and one that could stem from a simple laboratory accident."[40] Melting the world into goo, Drexler suggests, might not be the science fiction it seems at a time when rogues are looking for weapons of mass destruction.[41]

Nevertheless, Drexler takes us down the pro-nano path of Feyman's original vision. With the assistance of molecule-sized nanobots, Drexler says, we could construct material products one atom at a time. These machines could manufacture "anything that the laws of nature allow to exist." Indeed, we finally could turn base metals into gold. Or into fresh food, fresh air, even items that have no static nature, such as "clothing that becomes your bath water and then your bed." Nanosized robots could swarm through the bloodstream, repairing damaged cells, adjusting hair or skin color, and restoring lost youth.[42]

Could the dreams of the old alchemists finally become reality? Adam Keiper, writing in *The New Atlantis*, notes that so far no solid argument has been advanced that proves these things could not be done, though a number of scientists, including Nobel Prize–winner Richard Smalley, dismiss Drexler's vision as "just a dream," one that will "always remain a fantasy."[43] While no one yet knows how to make the molecular machines that could make these dreams a reality, neither has anyone found a fatal flaw in Drexler's theory. Dr. Bill Hurlbut describes nanotechnology as an "enabling technology" that "clearly is going to the bottom of the powers of matter, fundamental forces of nature to serve that which our human nature thinks nature ought to be, including revisions, potential revisions, of human nature" itself.[44]

Inventor Ray Kurzweil espouses Drexler's dream, particularly in its potential to change human nature. He expects that nanotechnology will be one among many technologies (such as artificial intelligence, robotics, and genetic engineering) that will irreversibly change human life. Kurzweil asks what will remain unequivocally human in a world in which physical reality is as mutable as virtual reality. His answer: "Ours is the species that inherently seeks to extend its physical and mental reach beyond current limitations."[45]

The twentieth-century theologian Reinhold Niebuhr spoke of this same human ambition, but he emphasized its perils. We are finite but strive to be infinite. Niebuhr, in *The Nature and Destiny of Man*, writes, "Man can find his true norm only in the character of God but is nevertheless a creature who cannot and must not aspire to be God."[46] This search to be like or to image the transcendent God, therefore, always includes a temptation to "sin," which Niebuhr defines as the "willful refusal to acknowledge the finite and determinate character of [human] existence."[47] If nanotechnology allows us to be godlike, then Niebuhr has already offered this warning against nano-hubris: "[Our] ability to stand outside and beyond the world tempts man to megalomania and persuades him to regard himself as the god around and about whom the universe centres. Yet he is too obviously involved in the flux and finiteness of nature to make such pretensions plausibly."[48]

Niebuhr wrote on the cusp of the revolution in atomic energy, but his principles surely apply to all technological advances. He urged human actions to solve problems, and the nanotechnology industry, by all reports, is moving ahead with safety and prudence in mind. But Niebuhr's point was also about the risk of taking wonder out of the world in our efforts to gain control. The wonder and meaning of life are in a source beyond human control and possession. He notes, "We can participate in the fulfillment of the meaning [of life] only if we do not seek too proudly to appropriate the meaning as our secure possession or to effect the fulfillment by our own power."[49]

When Martin Luther sat at his Wittenberg table discussing alchemy, he ended his comments by saying that changing one metal into another was an allegory for "the resurrection of the dead at the last day." Newton spent the last years of his life intensely studying alchemy and trying to decipher the timetables and secret numbers of the Bible in his own quest to come to terms with mortality. The quest for life everlasting continues in nanotechnology. Robert Freitas, a leading writer on nanomedicine, predicts that nanosized robots equipped to repair cells and deliver medications will give us the ability to halt aging and even reverse it:

> Once nanomachines are available, the ultimate dream of every healer, medicine man, and physician throughout recorded history will, at last, become a reality. Programmable and controllable microscale robots comprised of nanoscale parts fabricated to nanometer precision will allow medical doctors to execute curative and reconstructive procedures in the human body at the cellular and molecular levels. . . . [T]he ability to direct events in a controlled fashion at the cellular level is the key that will unlock the indefinite extension of human health and the expansion of human abilities.[50]

Today, hundreds of people have sought this indefinite extension of life by joining the extropian movement, which promises to preserve bodies into the future. Extropians have paid to have their bodies frozen in hopes of a nanotechnological resurrection at some future date. This is far from the resurrection promised in the Gospel, of course. The biblical resurrection is not about more time in our finite bodies. According to St. Paul, it is beyond time, a transcendence in which "we will all be changed" (1 Corinthians 15:51). The old alchemy was actually far more mystical than we can imagine today. Its advocates wrote about how its secrets and formulas changed hearts, as if into gold. Today's nanotechnology also reminds us how fundamentally changeable our world is, since nothing is static or immutable. Our reality is formed of atoms that are in continual motion. They form one thing now, but they could, and will, easily form something else in the future. The "solid" world is ultimately quite malleable. Both the universe and the human soul are dynamic, and this can surely evoke wonder.

Notes

1. Abdullah Suhrawardy, *The Sayings of Mohammad* (London: Constable, 1910), http://muslim-canada.org/hadiths.html.
2. John McCarthy, "Some Expert Systems Need Common Sense," *Annals of the New York Academy of Sciences* 426 (1984): 129–35.
3. Friedrich Nietzsche, *Thus Spake Zarathustra*, trans. Walter Kauffman (New York: Viking, 1954), 146.
4. John Haugeland, "What Is Mind Design," in *Mind Design II: Philosophy, Psychology, Artificial Intelligence*, ed. John Haugeland (Cambridge, MA: MIT, 1997), 26.
5. Benjamin Libet, "Do We Have Free Will?" *Journal of Consciousness Studies* 6, nos. 8–9 (1999): 47.
6. The main exception to this is the computer Dave in the film *2001: A Space Odyssey*.
7. For descriptions of a variety of mobile robots developed at MIT, see Rodney Brooks, "Elephants Don't Play Chess," *Robotics and Autonomous Systems* 6 (1990): 3–15.
8. Alan Turing, "Computing Machinery and Intelligence," in Haugeland, "What Is Mind Design," 29–32.
9. Ibid., 38.
10. While most in the AI community accept the Turing Test as sufficient, an opposing view can be found in John Searle, "Minds, Brains, and Programs," *Behavioral and Brain Sciences* 3 (1980): 417–24.
11. Terry Winograd and Fernando Flores, *Understanding Computers and Cognition: A New Foundation for Design* (Norwood, NJ: Ablex, 1986; reprint, Reading, MA: Addison-Wesley, 1991), 68.
12. J. Z. Young, *Programs of the Brain* (Oxford: Oxford University Press, 1978), 194.
13. Reported in Jonah Lehrer, "Hearts and Mind," *Boston Globe*, April 29, 2007.
14. Marvin Minsky, *The Society of Mind* (New York: Simon and Schuster, 1985).
15. Rosalind Picard, *Affective Computing* (Cambridge, MA: MIT Press, 1997), chapter 2.
16. Neal Stephenson, *Snow Crash* (New York: Bantam, 2000), 33.

17. Ray Kurzweil, *The Age of Spiritual Machines: When Computers Exceed Human Intelligence* (New York: Penguin, 1999), chapter 6.

18. Frank Tipler, *The Physics of Immortality: Modern Cosmology, God, and the Resurrection of the Dead* (New York: Doubleday, 1995).

19. Francis Crick, *The Astonishing Hypothesis: The Scientific Search for the Soul* (New York: Charles Scribner's Sons, 1994), 3.

20. Donald MacKay, *Behind the Eye* (Oxford: Basil Blackwell, 1991), 260.

21. Reinhold Niebuhr, *The Nature and Destiny of Man: A Christian Interpretation*, vol. 1, *Human Nature*, with intro. by Robin Lovin, Library of Theological Ethics (Louisville, KY: Westminster John Knox Press, 1996), 178–79.

22. Tipler, *Physics of Immortality*, 255. While this is not the place for an extended feminist critique, one cannot help but notice that the proponents of cybernetic immortality and artificial intelligence are overwhelmingly male. Women remain in their speculations as objects of desire yet are stripped of their reproductive role. Disembodied sexual experience, in the form of pornography, is, of course, a staple of the Internet.

23. Jaron Lanier, "One-Half of a Manifesto," *Wired* 8, no. 12 (December 2000): 4.

24. Joanna Bryson and Phil Kime, "Just Another Artifact," http://www.cs.bath. ac.uk/~jjb/web/aiethics98.html, accessed February 10, 2012.

25. Martin Luther, *Table Talk*, trans. William Hazlitt (Philadelphia: Lutheran Publication Society, 2004), 463.

26. Jacques Ellul, *The Technological Society* (New York: Knopf, 1964), 140–41.

27. Lynn White, "The Historical Roots of Our Ecological Crisis," *Science* 155 (March 10, 1967), 1203–7.

28. Robin Henig, "Our Silver-Coated Future," *OnEarth*, Fall 2007, 24.

29. Richard Feynman, "There's Plenty of Room at the Bottom," annual meeting of the American Physical Society, December 29, 1959, http://www.zyvex.com/ nanotech/feynman.html, accessed February 10, 2012.

30. Gabriel Silva, "Introduction to Nanotechnology and Its Applications to Medicine," *Surgical Neurology* 64 (2004): 216.

31. Prachi Patel-Predd, "Buckyballs with a Surprise," *MIT Technology Review* (November 1, 2006), http://www.technologyreview.com/read_article.aspx?id=17 681&ch=nanotech.

32. Buckyballs surpass diamonds and graphite in strength and hardness. Their inventors, Robert Curl, Richard Smalley, and Harold Kroto, received the Nobel Prize in chemistry in 1996.

33. Albert Fert and Peter Grundberg received the 2007 Nobel Prize in physics for this advance.

34. Henig, "Our Silver-Coated Future," 24.

35. James Thrall, "Nanotechnology and Medicine," *Radiology* 230 (2004): 317.

36. "Water-related Diseases," World Health Organization, http://www.who.int/ water_sanitation_health/diseases/diarrhoea/en/, accessed February 10, 2012.

37. Andrew Maynard outlined five difficulties for risk assessment to the President's Council on Bioethics. These included the difficulties of measuring exposure to nanomaterials, of testing these materials for toxicity, and of needing to look at the entire life cycle of nanoparticles, including their half-life in the environment. Transcript, "Nanotechnology: Benefits and Risks," President's Council

on Bioethics (July 23, 2007), 8.

38. Ibid., 19.

39. "Woodrow Wilson Center's Rejeski Explores How Nanotech May Affect Human Health," E and E TV, December 5, 2005, http://www.eande.tv/transc ripts/?date=120505#transcript, accessed February 10, 2012.

40. Bill Joy, "Why the Future Does Not Need Us," *Wired* 8.04 (April 2000), 61.

41. The development of "gray goo" would be extremely unlikely. Such a goo would need to be mobile, able to live off the biosphere (air or water), and self-replicating. Nature itself, with all the tools of evolution, has been unable to produce such a product. It is unlikely that human ingenuity would do so in the foreseeable future.

42. Eric Drexler, *Engines of Creation: The Coming Era of Nanotechnology* (New York: Anchor, 1986), reprinted at http://www.e-drexler.com/d/06/00/EOC/ EOC_Table_of_Contents.html.

43. Adam Keiper, "The Nanotechnology Revolution," *New Atlantis*, Summer 2003, 5, http://www.thenewatlantis.com/archive/2/keiperprint.htm, accessed February 10, 2012.

44. Hurlbut, President's Council on Bioethics, "Nanotechnology: Benefits and Risks," 12.

45. Ray Kurzweil, "The Next Frontier," *Science and Spirit*, November/December 2005, 69.

46. Niebuhr, *Nature and Destiny of Man*, 1:163.

47. 1:177.

48. 1:124–25.

49. 1:298.

50. Roert A. Frietas Jr., "Nanomedicine," http://www.foresight.org/Nanomedi cine/, accessed February 10, 2012.

Harold G. Koenig
Medicine, Religion, and Health

FROM AN ECONOMIC and policy point of view, Dr. Koenig offers no rosy vision of the health-care future. We will live longer, suffer more aging diseases, and need more medical facilities. On the other hand, his book *Medicine, Health, and Religion* argues that we have an untapped resource in people's religious beliefs, behaviors, ethics, and communities. To say such a thing in medicine a generation ago verged on heresy, Koenig says, but empirical studies are altering the view of clinicians. From a strictly scientific view, "Our information is still incomplete on the effects of religion and spirituality on mental and physical health," he says. In the gap, common sense has nevertheless alerted medicine that a majority of patients rely on religion in times of illness. To add hard evidence to this debate, Koenig's book surveys a vast number of clinical and population studies on health, disease, and religion. The studies looked at mind-body relations, mental health, immune systems, heart disease, stress, behavioral problems, longevity, and disability. Evidence is mounting that "religious beliefs influence coping with illness, affect medical decisions, and likely influenced medical outcomes." Religion and its communities will be a helpful resource amid the future health-care crisis. Even if scientific "proof" is elusive, research and the health-care crisis will make it "more and more difficult to ignore the spiritual needs of patients."

CHAPTER 9

Medicine Meets Modern Spirituality

Harold G. Koenig

WHENEVER I GIVE a public talk on religion and health issues, I try to avoid one thorny topic in particular: defining the differences between the words *religion* and *spirituality*. This can easily alienate a significant proportion of the audience because each of us has our own definitions for these words, which we hold onto quite dearly. Here, however, I have the luxury of spending some time exploring these terms in depth. Establishing definitions now for how I am using these terms will help the reader understand what the research means and will assist medical professionals in applying the findings to their clinical practices.

Without crystal-clear definitions, research on religion, spirituality, and health is not possible. For example, if a relationship is discovered between "spirituality" and longevity, what does that mean? The word *longevity* is widely understood as meaning years of life, and this can be calculated precisely by knowing birth and death dates. In contrast, there is no universal agreement on the more nebulous term *spirituality*. But if a relationship between spirituality and longevity is found, we need to know what this thing "spirituality" is in order to understand what exactly is related to a long life span.

We also need to know how spirituality differs from other psychosocial concepts, such as psychological well-being, altruism, forgiveness, humanism, social connectedness, and quality of life.

Spirituality must be unique and different from everything else, a completely separate phenomenon, which can then be examined in its relationship to health. Our task in conducting quantitative research is to quantify how spiritual the person is (determine the extent or degree that the person is spiritual) and describe in what ways he or she is spiritual. This is absolutely necessary in order to determine how spirituality is related to health.

To focus on these distinctions, this chapter compares four concepts—religion, spirituality, humanism, and positive psychology—with particular attention paid to spirituality, since this is such a commonly used term today. The meaning of the term *spirituality* has broadened in recent years to include positive psychological concepts such as meaning and purpose, connectedness, peacefulness, personal well-being, and happiness. According to researchers Christian Smith and Melinda Denton, "The very idea and language of 'spirituality,' originally grounded in the self-disciplining faith practices of religious believers, including ascetics and monks, then becomes detached from its moorings in historical religious traditions and is redefined in terms of subjective self-fulfillment."[1] This new version of spirituality has evolved to include not only aspects of life that have nothing to do with religion but often excludes religion entirely, as in the statement "I'm spiritual, not religious." This can make spirituality indistinguishable from concepts that are secular.

There are both positive and negative consequences to broadening of the term *spiritual*. I argue that we need to reinstate a sharper definition of *spirituality* that retains its historical grounding in religion. Nevertheless, I admit that the broadening of the term has a valuable clinical application. Spirituality can be profitably used in two different ways, more narrowly in research and more broadly for patient care. Before presenting a definition for *spirituality*, however, I first define *religion* and then review attempts by others to define spirituality as something unique and different from religion.

RELIGION

Religion may be defined as a system of beliefs and practices observed by a community, supported by rituals that acknowledge, worship, communicate with, or approach the Sacred, the Divine, God (in Western cultures), or Ultimate Truth, Reality, or nirvana (in Eastern cultures). Religion usually relies on a set of scriptures or teachings that describe the meaning and purpose of the world, the individual's place in it, the responsibilities of individuals to one another, and the nature of life after death. Religion typically offers a moral code of conduct that is agreed upon by members of the community, who attempt to adhere to that code.

Religious activity may be public, social, and institutional ("organizational" religiosity) or can be private, personal, and individual ("nonorganizational" religiosity). Organizational religiosity involves attending religious services, meeting as a group at other times for prayer or scripture study, or involvement with others in church-related activity, such as evangelism, fund-raising, financial giving, or church-related volunteering (I use the term *church* here for succinctness and ease of reading. What is intended is church, synagogue, mosque, or temple). Nonorganizational religiosity refers to religious activity that is done alone and in private, such as praying or communicating with God at home, meditating, reading religious scriptures, watching religious television or listening to religious radio, or performing private rituals such as lighting candles, wearing religious jewelry, and so forth.

Although religious practices (public and private) often reflect how deeply religious a person is, this is not always the case. To fill this gap, there is a dimension of religion that has to do with the importance or centrality of religion in life. This has been called *subjective religiosity* and is measured by researchers with self-ratings of religious importance or overall religiousness. There is also a motivational dimension of religion that is closely related to subjective religiosity but that asks *why* the person is religiously involved. What motivates the person to be religious? Is religion sought for its own value—that is, as an end in itself ("intrinsic" religiosity)—or is it being used as a means to some other end that is ultimately more important, such as social position or financial gain ("extrinsic" religiosity)? Although the organizational, nonorganizational, subjective, and motivational aspects of religion are considered by some to be the most salient, there are many other dimensions, including religious belief or orthodoxy, religious knowledge, religious coping, religious quest (or seeking), religious history, religious maturity, and religious well-being.[2]

In addition to its traditional form, religion also has a nontraditional shape, since it can be used to describe a wide array of groups guided by common beliefs and rituals. These include astrology, divination, witchcraft, invoking of spirits, spiritism, and a variety of indigenous, folk, or animistic rituals and practices related to the supernatural. Under my definition, therefore, religion is a unique domain with multiple dimensions that can be measured, quantified, and examined in relation to health and medical outcomes. Most high-quality research on religion, spirituality, and health actually ends up measuring religion, even many studies that use the term *spirituality* in the title or when discussing results.

SPIRITUALITY

I propose that, for pragmatic reasons, there should be two definitions of the term *spirituality*: one for conducting research and studying the relationship between spirituality and health, and one for applying what has been discovered to the care of patients. But first, how is *spirituality* usually defined? It has become a popular and accommodating term, especially in secular academic circles, because of its vagueness, broadness, and dependence on self-definition. This term can include everybody, even the nonreligious. Below are several definitions for the term by experts in the field.

The first three are definitions developed by health professionals involved in the clinical care of patients, which helps to explain why they are so broad and inclusive:

- The definition [of *spirituality*] is based in every person's inherent search for ultimate meaning and purpose in life. That meaning can be found in religion but may often be broader than that, including a relationship with a god/divine figure or transcendence; relationships with others; as well as spirituality found in nature, art, and rational thought.[3]
- The concept of spirituality is found in all cultures and societies. It is expressed in an individual's search for ultimate meaning through participation in religion and/or belief in God, family, naturalism, rationalism, humanism, and the arts. All of these factors can influence how patients and health care professionals perceive health and illness and how they interact with one another.[4]
- Spirituality is a complex and multidimensional part of the human experience. It has cognitive, experiential, and behavior aspects. The cognitive or philosophic aspects include the search for meaning, purpose, and truth in life and the beliefs and values by which an individual lives. The experiential and emotional aspects involve feelings of hope, love, connection, inner peace, comfort, and support. These are reflected in the quality of an individual's inner resources, the ability to give and receive spiritual love, and the types of relationships and connections that exist with self, the community, the environment and nature, and the transcendent (e.g., power greater than self, a value system, God, cosmic consciousness). The behavior aspects of spirituality involve the way a person externally manifests individual spiritual beliefs and inner spiritual state. Many people find spirituality through religion or through a personal relationship with the divine. However, others may find it through a connection to nature, through music and the

arts, through a set of values and principles or through a quest for scientific truth.[5]

Notice how these definitions of spirituality include meaning and purpose, inner peace and comfort, connection with others, support, feelings of wonder, awe, or love, and other healthy, positive terms. The definitions also make it very clear that spirituality does not have to involve religion—that is, it can be completely secular. Here, spirituality is defined by however a person chooses to define it, but it always means something good—something with which almost anyone would want to associate. While such broad definitions work well in clinical practice, they cause havoc when trying to conduct research. The difficulty of measuring such a concept and examining its relationship to health, especially mental health, should be obvious.[6] I say more about this later.

Peter C. Hill and Ken Pargament, well-known researchers in this area, define *spirituality* in a more unique way to help distinguish it from other related concepts. They argue that in the United States and elsewhere there is

a polarization of religiousness and spirituality, with the former representing an institutional, formal, outward, doctrinal, authoritarian, inhibiting expression and the latter representing an individual, subjective, emotional, inward, unsystematic, freeing expression. . . . [S]pirituality can be understood as a search for the sacred, a process through which people seek to discover, hold on to, and, when necessary, transform whatever they hold sacred in their lives. . . . This search takes place in a larger religious context, one that may be traditional or non-traditional. . . . The sacred is what distinguishes religion and spirituality from other phenomena. It refers to those special objects or events set apart from the ordinary and thus deserving of veneration. The sacred includes concepts of

God, the divine, Ultimate Reality, and the transcendent,
as well as any aspect of life that takes on extraordinary
character by virtue of its association with or representation of such concepts. . . .[7]

In another article, Pargament further explains:

I see spirituality as a search for the sacred. It is, I believe, the
most central function of religion. It has to do with however people think, feel, act, or interrelate in their efforts
to find, conserve, and if necessary, transform the sacred
in their lives. Let me say a bit more about the sacred. In
the *Oxford English Dictionary*, the sacred is defined as
the holy, those things "set apart" from the ordinary, worthy of reverence. The sacred encompasses concepts of
God, the divine, the transcendent, but it is not limited
to notions of higher powers. It also includes objects,
attributes, or qualities that become sanctified by virtue
of their association with or representation of the holy.[8]

David J. Hufford, an authority in the medical humanities who
holds a PhD in folklore and folk life from the University of Penn-
sylvania, makes the following observation:

The odd thing about the inconsistency, vagueness and
worry by investigators over these terms [religion and spir-
ituality] is that they do have consistent, concise meanings in ordinary speech, and they relate to one another in
a perfectly ordinary way. We are accustomed to pairs of
words such as learning and education or health and medicine, where the former word identifies a broad domain
and the second word refers to an institutional aspect of
that domain. . . . Not all learning happens in schools and
not all health behavior takes place in clinics or hospitals.

> Spirituality and religion stand in the same relation. Spir-
> ituality refers to the domain of spirit(s): God or gods,
> souls, angels, kjinni, demons. In short, this is what was
> once called the supernatural (and still is by many English
> speakers). When spirituality refers to something else it is
> by metaphorical extension to other intangible and invis-
> ible things. . . .[9]

In attempting to generalize his observations, Hufford specifically comments on the applicability of what he is saying to Eastern traditions.

> It is sometimes suggested that spirit(s) comprise a West-
> ern category and that some traditions, Buddhism being
> an often cited example, lack the concept. But as long as
> the concept is kept simple in definition this is not a valid
> criticism. The concept of reincarnation in Buddhism
> may not involve a concept analogous to the Western idea
> of a soul in some of its versions, but it nonetheless does
> involve something invisible and intangible that is a kind
> of essence of the person that reincarnates. . . . It is some-
> times claimed that Buddhism is not a religion, some-
> times defended on the basis that it is not theistic. Even
> apart from the fact that much of Buddhist beliefs and
> practice around the world DOES involve gods, clearly
> Buddhism is an institution organized around such ideas
> as reincarnation and Nirvana and it teaches practices
> that affect the intangible part of the human, the part that
> progresses or degenerates, that approaches enlighten-
> ment and Nirvana.[10]

In the end, Hufford defines *spirituality* simply as "personal relationship to the transcendent" and *religion* as "the community, institutional aspects of spirituality."[11]

RECOMMENDED DEFINITION

For research purposes, my definition of *spirituality* comes closest to Hufford's. To facilitate measurement as a unique and distinct concept, I believe we should return the definition of *spirituality* to its origins in religion, whether traditional or nontraditional.[12] In its historical usage, the term *spirituality* has its roots in the patristic era and later spiritual thought derived from monastic life, the mendicants, the late Middle Ages, Luther, Ignatius, Teresa, and John of the Cross. Because the word *spirituality* has historically been associated with religion or the supernatural and involves religious language, I argue that, to call something "spiritual," it must have some connection to religion.

Bear in mind that my definition of *religion* above includes non-traditional religious expressions. These include astrology, divination, witchcraft, folk traditions, or other indigenous healing practices that involve invisible spirits and spiritual forces that are outside of the individual yet are often practiced with others as part of a community with a set of beliefs, rituals, and moral code. My definition of the word *religion* also includes personal and private kinds of beliefs and activities not tied to organized or institutional worship. Religion also includes searching for or seeking the Sacred or transcendent, as the religious quest dimension would measure. However, if there is no connection with either religion or the supernatural, then I would not call a belief, practice, or experience "spiritual." I would call it "humanistic." An apt definition of the latter would be the following:

> Humanism is a broad category of active ethical philosophies that affirm the dignity and worth of all people, based on the ability to determine right and wrong by appeal to universal human qualities—particularly rationalism. Humanism is a component of a variety of more specific philosophical systems, and is also incorporated

into some religious schools of thought. Humanism entails a commitment to the search for truth and morality through human means in support of human interests. In focusing on the capacity for self-determination, Humanism rejects transcendental justifications, such as a dependence on faith, the supernatural, or divinely revealed texts.[13]

In the area of spirituality, another important distinction must be made. When someone speaks of inner peace; connection with others; purpose and meaning; beliefs and values; feelings of wonder, awe, or love; forgiveness; gratitude; comfort; support; and other quasi-indicators of good mental health, this is not spirituality itself but the *result* of devoutly practiced spirituality. These good things are the consequences of living a spiritual life, not spirituality itself. Whether spiritual sources are more likely than secular sources to result in these positive psychological states is something that research must determine; spirituality cannot be defined by its consequences. Simply defining spirituality as good mental heath and including mental health indicators as part of the measures of spirituality precludes any ability to actually study the relationship between spirituality and mental health. To do so with a measure of spirituality contaminated by questions tapping positive psychological states or traits is called "tautology," a kind of circular reasoning that involves correlating a concept with itself. This kind of research always finds a positive correlation between spirituality and mental health (and often with physical health as well, due to the close connection between mental and physical health). Thus, when spirituality is used in research to study health, it cannot be defined in terms of positive or healthy psychological or social states.[14]

In the care of patients, however, it is not necessary to define *spirituality* as rigorously as when conducting scientific research. In clinical settings, it is more useful to define *spirituality* as broadly as possible so that all patients have an opportunity to have their spiritual needs addressed (in whatever way they define those spiritual needs). Some patients may not view themselves as religious but may nevertheless be searching for greater meaning outside themselves or struggling with existential concerns. Those may not be neatly categorized into psychological or social problems that lie within the expertise of mental health professionals or social workers. Again, definitions of *religion* and *spirituality* do not have to be as crisp and distinctive in clinical environments as they do in research settings.

Many patients may not understand the differences that academic health professionals are now making between religion and spirituality. In research that asked 838 patients to categorize themselves as either religious, spiritual, both, or neither, close to 90 percent indicated that they were both religious and spiritual.[15] Thus, a term is needed that everyone—religious and nonreligious alike— can relate to. It is not surprising, then, that definitions of *spirituality* by health professionals that I describe above are so broad and inclusive. But this will not do in research, which seeks to be discriminating, exclusive, and reductionistic in order to determine exactly what is affecting what.

Having different definitions for *spirituality* depending on the setting (research versus clinical) is not without potential problems, however. What if research shows that religious types of spirituality are related to better health, whereas nonreligious spirituality is either unrelated to health or related to worse health outcomes? In studies performed thus far, the finding is that greater *religious* involvement is related to better health—with the research strongest for mental health and less solid (but still impressive) for physical health. Other studies show that most of the existing research lumps "spiritual-but-not-religious" patients into the nonreligious group, the one that appears to be doing worse than religious patients. These studies seldom examine the spiritual-but-not-religious patients separately from other nonreligious patients. But the idea that spiritual-but-not-religious types may actually do worse is still a distinct possibility. What then? Should clinicians be asked to support the beliefs of spiritual-but-not-religious patients whose belief system may be having no effect or possibly even negative effects on health?

I believe the overall goal of the clinician is to find common ground with all patients, which means not trying to change beliefs, but rather trying to support beliefs that help patients cope. Using *spirituality* in its broadest definition, then, makes sense in clinical practice.

From Definitions to Practice

Finding the proper distinctions between religion, spirituality, humanism, and other psychosocial concepts is especially important for research that wants to identify the exact causes of better health and medical outcomes. For clinical purposes, however, it is probably best to use a broader definition of *spirituality* that includes religious and nonreligious types and is self-defined by patients themselves. We want as many patients as possible to have an opportunity to have their spiritual needs identified and addressed, however they understand them.

Even in the twenty-first century, patients seem eager during times of illness to speak of their concerns and fears in personal religious terms. In our hospitals and clinics, it is not uncommon to find patients who will confide a story like this:

> Doctor, you say that I have terminal cancer and there isn't any more that you can do for me. You say that I have two or three months left. What happens then? I'm afraid of the pain and suffering ahead. I'm afraid that I haven't been a good person. I'm afraid that God doesn't love me, since my prayers for healing have gone unanswered. I'm afraid of where I'm going after I die. I'm afraid of leaving

my daughter and son, and never seeing them again. I'm afraid, doctor, I'm so afraid.

Increasingly, health professionals and the medical system must tackle such personal inquiries. The questions for medical practice abound. Should health professionals take more seriously the spiritual concerns of patients, and can this be a way to approach individual patients more compassionately as well as to strengthen and reform the health-care system? If these spiritual concerns are taken into account, what might health care look like thirty years from now? If health-care systems in the coming decades can no longer function as they have in the past, what are the options and how might faith communities be helpful?

Our approach to such policy questions will help determine the quality of care that patients receive in the future. The good news is that research on religion, spirituality, and health is advancing as never before, even though most doctors are still not trained to talk to patients about these issues. The bad news is in the headlines every day: the health-care systems of the world are headed for troubled times, unless we find innovative solutions.

RELIGION-HEALTH RESEARCH

By the year 2000, the number of studies examining the relationship between some aspect of religion, spirituality, and health or health care had ballooned to nearly 1,200 (about 70 percent were on mental health and 30 percent on physical health).[16] Since then, nearly 2,000 more studies have been published. Thus, it is safe to say that nearly 3,000 research studies have quantitatively examined relationships between religion, spirituality, and health, many reporting positive findings.

This research has prompted at least three consensus conferences partially supported by the National Institutes of Health (NIH) to review the research and come up with recommendations for methodological advances and future studies.[17] William R. Miller, chair of the latest NIH working group, concluded, "Substantial empirical evidence points to links between spiritual/religious factors and health in U.S. populations."[18] While it is clear that many of the hundreds of studies published in the scientific literature have serious methodological flaws, not all of the studies do, and the critique of earlier research may have been overstated.[19] Furthermore, the quality of religion-health studies has increased substantially since the last NIH conference in 2002, with investigators responding to and correcting many of the concerns voiced about prior research.

While the field of religion, spirituality, and health is in its infancy and much research is needed to verify (or dispel) earlier findings, a lot of outstanding work has already been done. There is good reason to begin implementing some of what is already known in clinical practice.

CLINICAL PRACTICE APPLICATIONS

Not all of the reasons for addressing spiritual issues in clinical practice depend on research that definitively demonstrates that religion influences health. The application is for very practical reasons: many patients are religious, have religious beliefs and traditions related to health, and have health problems that often give rise to spiritual needs. Religious beliefs will frequently influence the kind of health care that patients wish to receive. Those beliefs affect how patients cope with illness and derive meaning and purpose when feeling bad physically or unable to do the things they used to do that give them joy and pleasure. Such beliefs help patients maintain hope and motivation toward self-care in the midst of overwhelming circumstances. Patients, particularly when hospitalized, may be isolated from their religious communities, and, because spiritual needs often come up during this time, health-care providers must recognize and address those needs. Religious beliefs can also influence medical decisions, conflict with medical treatments, and influence patients' compliance with treatments prescribed. The patient's involvement in a religious community can affect the support and monitoring he or she receives after discharge. In summary, there are many reasons for clinicians to discuss religious or spiritual issues with patients, learn to identify spiritual needs, and refer patients to health professionals trained to address those needs.[20]

The need for training to integrate spirituality into patient care has been increasingly recognized within medical education. In 1992, only 3 medical schools offered courses on religion, spirituality, and medicine. By 2006, over 100 of the 141 medical schools in the United States and Canada had such courses (70 percent of which are required).[21] Unfortunately, most physicians in practice today have no such training. In a recent nationwide survey of a random sample of over one thousand U.S. physicians of all specialties, only about 10 percent of physicians indicated they routinely talked to patients about these issues.[22] Those data are consistent with reports by patients. Only 10 to 20 percent of patients report that a physician *ever* asked about spiritual issues.[23] As leaders in health care, physicians ought to be responsible for ensuring that spiritual needs likely to affect medical decisions and health outcomes are addressed.

Evidence that even the spiritual needs of dying patients are often unmet and the adverse effect of this on quality of life has recently been reported. Balboni and colleagues surveyed 230 patients with advanced cancer who had failed to respond to first-line chemotherapy.[24] These patients were being cared for at some of the best medical-care systems in the world, including Yale University Cancer Center in New Haven, Connecticut, and Memorial Sloan-Kettering Cancer Center in New York City. This study, conducted by Harvard Medical School researchers, had patients rate on a 1–5 scale to what extent either their religious community or the medical system supported their spiritual needs (from "not at all" to "completely supported"). One out of every two patients (47 percent) said that spiritual needs were minimally or not at all met by their religious community. Nearly three-quarters (72 percent) said that spiritual needs were minimally or not at all met by the medical system (i.e., doctors, nurses, or chaplains). Patients who indicated that their spiritual needs were being met reported significantly higher quality of life. In fact, of nine factors known to influence quality of life, degree of spiritual support was the second-strongest predictor.

Unfortunately, there are not enough chaplains employed by hospital organizations to screen all patients or address the spiritual needs that are present, nor do community clergy have the time or expertise to meet those needs. Chaplains see only about 20 percent of hospitalized patients in the United States today.[25] In the current environment of intense competition among hospitals to survive financially, chaplain services are often the first to be downsized or eliminated.[26] The results of a survey on patient satisfaction that involved 1,732,562 patients representing 33 percent of all hospitals in the United States and 44 percent of all hospitals with more than one hundred beds found that satisfaction with the emotional and spiritual aspects of care received one of the lowest ratings of all clinical-care indicators. At the same time, it was one of the areas rated the highest for need of quality improvement.[27] This is a major reason why physicians, nurses, social workers, and other health professionals need to get more involved. There are not enough chaplains in hospital settings to see all the patients, so health professionals need to identify patients with the most pressing spiritual needs and get them connected to the few chaplains who are available to address them. However, there is resistance to doing so, particularly among physicians.

Physician Attitudes

Despite what we know about the spiritual needs of patients and their relationship to health and well-being, few health professionals are addressing them. Most of what we know about the behaviors and attitudes of health professionals comes from studies of physicians. As noted earlier, only one in ten doctors regularly addresses spiritual issues with patients. So how do physicians in general feel about becoming more involved in this area?

Most physicians (over 90 percent) acknowledge that spiritual factors are an important component of health, and the majority (70 to 82 percent) say that this can influence the health of the patient.[28] Furthermore, 85 percent of physicians say that they should be aware of the patient's religious/spiritual beliefs, and 89 percent indicate that they have a right to inquire about those beliefs.[29] Despite these positive attitudes, however, physicians are reluctant to talk with patients about spiritual issues or to take a spiritual history.

Doctors respond differently to such questions when asked about patients being treated in different settings, such as in outpatient clinics, acute hospital environments, or hospice-type circumstances. Depending on setting, 31 to 74 percent of physicians feel that they should take a spiritual history. They are more likely to do this as the patient's medical condition becomes more severe.[30] The best data on physicians' attitudes toward taking a spiritual history come from Curlin and colleagues' national survey of physicians mentioned earlier (see note 22 in this chapter). In that study, 55 percent said it was usually or always appropriate for physicians to inquire about patients' religious/spiritual beliefs, whereas 45 percent said it was usually or always *inappropriate* to do so. Thus, there appears to be a gap between what physicians feel they need to know and what they feel is appropriate to do in order to gather that information.

Interestingly, as the Curlin and colleagues survey shows, the strongest predictor of whether a physician addresses spiritual issues with patients has nothing to do with the patient's condition or the patient's interest in spirituality. Rather, it depends on how religious or spiritual the physician is. Common sense in this era of patient-centered medicine dictates that it ought to be the characteristics of the patient that determine whether spiritual matters are addressed, not the characteristics of the physician. Having insufficient time is also not the most common reason for ignoring patients' spiritual needs.[31] Moreover, recent data, again supplied by Curlin and associates, suggest that physicians who say that time is a barrier to addressing spiritual needs are actually more likely to talk with patients about these issues than physicians who don't indicate that time is a problem.

In my opinion, physicians should *make* time to inquire about spiritual issues that may directly or indirectly influence the health and health care of patients. A brief screening spiritual history takes only a few minutes. Physicians may choose to do other interventions beyond simple inquiry, although they will depend on the comfort level of the physician. Such interventions involve supporting the religious/spiritual beliefs of the patient and, if requested, praying with patients. In the national survey of physicians by Curlin and colleagues, 73 percent of physicians said that they often or always encouraged the patient's own religious/spiritual beliefs and practices (versus sharing their own religious beliefs with patients). Other research (now twenty years old) suggests that about one-third of physicians have at some point in their careers prayed with a patient, with the vast majority of these physicians reporting that it had benefited the patient.[32] While many physicians feel that it is appropriate for them to pray with patients *if the patient asks*, only a small percentage (6 to 30 percent) say that it is appropriate for physicians to initiate prayer with patients.[33] Most physicians feel uncomfortable about initiating prayer with patients, particularly physicians who are not religious themselves. In general, though, physicians feel that the sicker the patient is, the more appropriate it is to inquire about spiritual matters or pray with him or her.

In summary, most physicians (and probably other health professionals as well) recognize the importance and value of patients' spiritual beliefs in the health of those patients and feel that they need to know about these beliefs. However, when asked specifically about what they are doing and what is appropriate to do, few physicians actively assess or address spiritual issues or are open to doing more in this area. Several forces on the horizon, though, are likely to change the attitudes of health-care professionals and health-care systems toward addressing spiritual issues and working with religious communities.

Notes

1. C. Smith and M. L. Denton, *Soul Searching: The Religious and Spiritual Lives of American Teenagers* (New York: Oxford University Press, 2005), 175.
2. A. Futterman and H. G. Koenig, "Measuring Religiosity in Later Life: What Can Gerontology Learn from the Sociology and Psychology of Religion?, background paper, published in *Proceedings of Conference on Methodological Approaches to the Study of Religion, Aging, and Health,* cosponsored by NIA and Fetzer Institute, March 16–17, 1995.
3. C. M. Puchalski, "Spirituality and Medicine: Curricula in Medical Education," *Journal of Cancer Education* 21 (2006): 14–15.
4. Association of American Medical Colleges, "Contemporary Issues in Medicine: Communication in Medicine," Medical School Objectives Project, Report III, 1999. Available at http://www.aamc.org/meded/msop/msop3.pdf (accessed January 2007).
5. G. Anandarajah and E. Hight, "Spirituality and Medical Practice: Using the HOPE Questions as a Practical Tool for Spiritual Assessment," *American Family Physician* 63 (2001): 83.
6. A. Moreira-Almeida and H. G. Koenig, "Retaining the Meaning of the Words Religiousness and Spirituality: A Commentary on the WHOQOL SRPB Group's 'A Cross-Cultural Study of Spirituality, Religion, and Personal Beliefs as Components of Quality of Life,'" *Social Science and Medicine* 63 (2006): 843–45.
7. P. C. Hill and K. I. Pargament, "Advances in the Conceptualization and Measurement of Religion and Spirituality," *American Psychologist* 58 (2003): 64–65.
8. K. I. Pargament, "The Psychology of Religion *and* Spirituality? Yes and No," *International Journal for the Psychology of Religion* 9 (1999): 12.
9. D. J. Hufford, "An Analysis of the Field of Spirituality, Religion and Health. Area I Field Analysis," 2005, http://www.metanexus.net/tarp/pdf/TARP-Hufford.pdf, accessed January 2007.
10. Ibid.
11. Ibid.
12. H. G. Koenig, "Religion, Spirituality and Medicine in Australia: Research and Clinical Practice," *Medical Journal of Australia* 186 (2007): S45–S46.
13. Wikipedia, s.v. "Humanism," http://en.wikipedia.org/wiki/Humanism.
14. H. G. Koenig. "Concerns about Measuring 'Spirituality' in Research," *Journal of Nervous and Mental Disease,* 196, no. 5 (2008): 349–55.

15. H. G. Koenig, L. K. George, and P. Titus, "Religion, Spirituality and Health in Medically Ill Hospitalized Older Patients," *Journal of the American Geriatrics Association* 52 (2004): 554–62.

16. H. G. Koenig, M. M. McCullough, and D. B. Larson, *Handbook of Religion and Health* (New York: Oxford University Press, 2001), 514–89.

17. For information on the recommendations of these conferences, see: National Institute on Aging and Fetzer Institute, Conference on Methodological Approaches to the Study of Religion, Aging, and Health, Bethesda, MD, 1995; National Institute on Aging and Fetzer Institute, Working Group on Measurement of Religion/Spirituality for Healthcare Research, Bethesda, MD, 1997 (in particular, see E. L. Idler, M. A. Musick, C. G. Ellison, et al., "Measuring Multiple Dimensions of Religion and Spirituality for Health Research: Conceptual Background and Findings from the 1998 General Social Survey," *Research on Aging* 25 [2003]: 327–65 and National Institutes of Health Working Group on Spirituality, Religion, and Health, Office of Behavioral and Social Sciences Research, Bethesda, MD, 2001. This last report resulted in a special section on spirituality, religion, and health published in *American Psychologist* 58, no. 1 [January 2003].

18. W. R. Miller and C. E. Thorsen, "Spirituality, Religion and Health: An Emerging Research Field," *American Psychologist* 38 (2003): 33.

19. For questions on methodology, see R. P. Sloan, E. Bagiella, and T. Powell, "Religion, Spirituality, and Medicine," *Lancet* 353 (1999): 664–67. For a rebuttal, see H. G. Koenig, E. Idler, S. Kasl, et al., "Religion, Spirituality, and Medicine: A Rebuttal to Skeptics," *International Journal of Psychiatry in Medicine* 29 (1999): 123–31.

20. H. G. Koenig, *Spirituality in Patient Care*, 2nd ed. (Philadelphia: Templeton Foundation Press, 2007).

21. Regarding these statistics, see C. M. Puchalski, "Spirituality and Medicine: Curricula in Medical Education," *Journal of Cancer Education* 21, no. 1 (2006): 14–18; and *John Templeton Foundation Capabilities Report* (West Conshohocken, PA: Templeton Foundation, 2006), 68.

22. F. A. Curlin, M. H. Chin, S. A. Sellergren, C. J. Roach, and J. D. Lantos, "The Association of Physicians' Religious Characteristics with Their Attitudes and Self-Reported Behaviors Regarding Religion and Spirituality in the Clinical Encounter," *Medical Care* 44 (2006): 446–53.

23. See T. McNichol, "The New Faith in Medicine," *USA Weekend*, April 5–7, 1996, 5; J. Ehman, B. Ott, T. Short, R. Ciampa, and J. Hansen-Flaschen, "Do Patients Want Physicians to Inquire about Their Spiritual or Religious Beliefs If They Become Gravely Ill?" *Archives of Internal Medicine* 159 (1999): 1803–6; and D. E. King and B. Bushwick, "Beliefs and Attitudes of Hospital Inpatients about Faith Healing and Prayer," *Journal of Family Practice* 39 (1994): 349–52.

24. T. A. Balboni, L. C. Vanderwerker, S. D. Block, et al., "Religiousness and Spiritual Support among Advanced Cancer Patients and Associations with End-of-Life Treatment Preferences and Quality of Life," *Journal of Clinical Oncology* 25 (2007): 555–60.

25. K. J. Flannelly, K. Galek, and G. F. Handzo, "To What Extent Are the Spiritual Needs of Hospital Patients Being Met?" *International Journal of Psychiatry in*

Medicine 35, no. 3 (2005): 319–23.

26. L. VandeCreek, "How Has Health Care Reform Affected Professional Chaplaincy Programs and How Are Department Directors Responding?" *Journal of Health Care Chaplaincy* 10, no. 1 (2000): 7–17.

27. P. A. Clark, M. Drain, and M. P. Malone, "Addressing Patients' Emotional and Spiritual Needs," *Joint Commission Journal on Quality and Safety* 29 (2003): 659–70.

28. On physicians' acknowledging the importance of spirituality, see M. R. Ellis, D. C. Vinson, and B. Ewigman, "Addressing Spiritual Concerns of Patients: Family Physicians' Attitudes and Practices," *Journal of Family Practice* 48 (1999): 105–9. On spirituality's influencing health, see H. G. Koenig, L. Bearon, and R. Dayringer, "Physician Perspectives on the Role of Religion in the Physician–Older Patient Relationship," *Journal of Family Practice* 28 (1989): 441–48; and C. A. Armbruster, J. T. Chibnall, and S. Legett, "Pediatrician Beliefs about Spirituality and Religion in Medicine: Associations with Clinical Practice," *Pediatrics* 111 (2003): E227–E235.

29. See M. H. Monroe, D. Bynum, B. Susi, et al., "Primary Care Physician Preferences Regarding Spiritual Behavior in Medical Practice," *Archives of Internal Medicine* 163 (2003): 2751–56; and T. A. Maugans and W. C. Wadland, "Religion and Family Medicine: A Survey of Physicians and Patients," *Journal of Family Practice* 32 (1991): 210–13.

30. Monroe, Bynum, Susi, et al., "Primary Care Physician Preferences Regarding Spiritual Behavior in Medical Practice."

31. J. T. Chibnall and C. A. Brooks, "Religion in the Clinic: The Role of Physician Beliefs," *Southern Medical Journal* 94 (2001): 374–79.

32. Koenig, Bearon, and Dayringer, "Physician Perspectives on the Role of Religion."

33. See Monroe, Bynum, Susi, et al., "Primary Care Physician Preferences Regarding Spiritual Behavior in Medical Practice"; and Curlin, Chin, Sellergren, Roach, and Lantos, "Association of Physicians' Religious Characteristics."

Acknowledgments

TEMPLETON PRESS published nine books in the Templeton Science and Religion Series from 2008 through 2011. Every chapter in this volume originally appeared in one of those volumes.

Chapter 1: "Case for the Big Bang," adapted from Joseph Silk, *Horizons of Cosmology: Exploring Worlds Seen and Unseen,* 2009.

Chapter 2: "Rocks, Time, Fossils, and Life Itself," adapted from Ian Tattersall, *Paleontology: A Brief History of Life,* 2010.

Chapter 3: "From Deluge to Biogeography," adapted from R. J. Berry, *Ecology and the Environment: The Mechanisms, Marring, and Maintenance of Nature,* 2001.

Chapter 4: "The Human Primate: A Quantum Leap?" adapted from Malcolm Jeeves and Warren S. Brown, *Neuroscience, Psychology, and Religion: Illusions, Delusions, and Realities about Human Nature,* 2009.

Chapter 5: "How Genetics Rescued Darwinian Evolution," adapted from Denis R. Alexander, *The Language of Genetics: An Introduction,* 2011.

Chapter 6: "How We Conceive of the Divine," adapted from Justin L. Barrett, *Cognitive Science, Religion, and Theology: From Human Minds to Divine Minds,* 2011.

Chapter 7: "On Math and Metaphysical Language," adapted from Javier Leach, *Mathematics and Religion: Our Language of Sign and Symbol,* 2010.

Chapter 8: "Between Cyperspace and the New Alchemy," adapted from Noreen Herzfeld, *Technology and Religion: Remaining Human in a Co-Created World*, 2009.

Chapter 9: "Medicine Meets Modern Spirituality," adapted from Harold G. Koenig, MD, *Medicine, Health, and Religion: Where Science and Spirituality Meet*, 2008.

 Contributors

Denis Alexander is director of the Faraday Institute for Science and Religion at St. Edmund's College, Cambridge, where he is a fellow. A biochemist by background, Dr. Alexander has spent the past forty years in the biological research community, most recently at the Babraham Institute, Cambridge.

Justin L. Barrett is Thrive Professor of Developmental Science and director of the Thrive Center for Human Development at Fuller Theological Seminary's School of Psychology. He is author of numerous articles concerning cognitive, developmental, and evolutionary approaches to the study of religion. His authored books are *Why Would Anyone Believe in God?*, *Cognitive Science, Religion, and Theology: From Human Minds to Divine Minds*, and *Born Believers: The Science of Childhood Religion*. He edited a four-volume collection, *Psychology of Religion*.

R. J. Berry was professor of genetics at University College London from 1974–2000. He is a Gifford Lecturer, a former president of the British Ecological Society and of the European Ecological Federation, and a founding member of the International Society for Science and Religion.

Warren S. Brown is professor of psychology and director of the Travis Research Institute at the Fuller Graduate School of Psychology. He is a research neuropsychologist with more than eighty published scientific publications and is the coauthor of several books: *Whatever Happened to the Soul?*; *Did My Neurons Make Me Do It?*; *Neuroscience, Psychology and Religion*; and *Physical Nature of Christian Life*.

Noreen Herzfeld is the Nicholas and Bernice Reuter Professor of Science and Religion at St. John's University, Collegeville, Minnesota. Her publications include *In Our Image: Artificial Intelligence and the Human Spirit, Technology and Religion: Remaining Human in a Co-Created World*, and *The Limits of Perfection*.

Malcolm Jeeves is emeritus professor of psychology at St. Andrews University and formerly editor-in-chief of the leading neuroscience journal *Neuropsychologia* . He is a past president of the Royal Society of Edinburgh, Scotland's National Academy. He is author of *Human Nature* and editor and contributor to *Rethinking Human Nature* and *From Cells to Souls—and Beyond*.

Harold G. Koenig, MD, is on the faculty at Duke University as professor of psychiatry and behavioral sciences and associate professor of medicine. Dr. Koenig is also director of the Center for Spirituality, Theology, and Health at Duke University Medical Center and is Distinguished Adjunct Professor at King Abdulaziz University in Jeddah, Saudi Arabia. Dr. Koenig has published extensively in the fields of mental health, geriatrics, and religion, with over 350 scientific peer-reviewed articles and book chapters and nearly 40 books in print or in preparation.

Javier Leach has degrees in philosophy, theology, and mathematics and is also a Jesuit priest. Currently he teaches logic and mathematics at the Complutense University of Madrid and advises the chair of science, technology, and religion at the Comillas University in Madrid, which he directed from 2003 to 2011.

Joseph Silk is professor at the Institut d'Astrophysique and Pierre and Marie Curie University, Paris, and also holds positions at Johns Hopkins University and the University of Oxford. Most of his scientific research is related to cosmology and particle astrophysics, and he has published more than seven hundred articles and several popular books.

Ian Tattersall is a curator emeritus in the Division of the American Museum of Natural History, New York City. In paleontology, his principal research interests are in human evolution, notably in species recognition in the fossil record and in the origin of modern human cognition.

 Index